POLICE LEADERSHIP

Second Edition

POLICE LEADERSHIP

ORGANIZATIONAL AND MANAGERIAL
DECISION MAKING PROCESS

M. R. Haberfeld, Ph.D.
John Jay College of Criminal Justice

PEARSON

Boston   Columbus   Indianapolis   New York   San Francisco   Upper Saddle River
Amsterdam   Cape Town   Dubai   London   Madrid   Milan Munich   Paris   Montreal   Toronto
Delhi   Mexico City   Sao Paulo   Sydney   Hong Kong   Seoul   Singapore   Taipei   Tokyo

Editorial Director: Vernon Anthony
Senior Acquisitions Editor: Eric Krassow
Assistant Editor: Tiffany Bitzel
Editorial Assistant: Lynda Cramer
Director of Marketing: David Gessell
Senior Marketing Manager: Cyndi Eller
Senior Marketing Coordinator: Alicia Wozniak
Senior Marketing Assistant: Les Roberts

Production Manager: Holly Shufeldt
Senior Art Director: Jayne Conte
Cover Designer: Karen Noferi
Cover Image: Dreamstime
Full-Service Project Management: Dhanya Ramesh, Jouve
Composition: Jouve
Text and Cover Printer/Binder: LSC Communications, Inc.

Credits and acknowledgments borrowed from other sources and reproduced, with permission, in this textbook appear on the appropriate page within text.

Library of Congress Cataloging-in-Publication Data

Haberfeld, M. R. (Maria R.)
  Police leadership: organizational and managerial decision making process / M. R. Haberfeld. —2nd ed.
    p. cm.
  Includes bibliographical references and index.
  ISBN 978-0-13-268296-1
   1. Police—Supervision of.   2. Leadership.   3. Interpersonal relations.   I. Title.
  HV7936.S8H36 2013
  363.2068'4—dc23

                                          2011037988

ISBN 10: 0-13-268296-6
ISBN 13: 978-0-13-268296-1

To the memory of my beloved grandfather,
Jakub Rudenski—you were the best proof that water
can be as thick as blood.

CONTENTS

Preface   xi
Acknowledgments   xiii

Chapter 1   INTRODUCTION: THE PENTAGON OF POLICE LEADERSHIP   1
The Case for Early Career Leadership Training   1
The Training Gap   3
The Pentagon of Police Leadership   5

Chapter 2   INTEGRITY, ETHICS, AND POLICE LEADERSHIP   12
Ethics and Transformational Leadership   13
Integration of Behavior and Values   14
Stewardship Model   15
Ethical Leaders' Characteristics   15
The Triangle of Police Integrity   16
The Pentagon of Police Leadership   19

Chapter 3   PARTNERSHIP IN A SMALL FORCE: TEAM THEORY   23
Overview of the Theory   23
Functional Team Leadership   23
The Leader-Management Model   25
Self-Managed Teams   26
Team Policing   26

Chapter 4   IN-GROUPS AND COMMUNITY-ORIENTED POLICING: LEADER–MEMBER EXCHANGE THEORY   33
Overview of the Theory   33
Measurement of the Quality of the Dyadic Relationship   34
The Dyadic Relationship and Effectiveness   34
Antecedents to High-Quality Dyadic Relationships   36
Leader–Member Exchange Theory and Bias   36
LMX Theory and the Cases of Commissioner Lee P. Brown and Chief Darrel Stephens   37

Chapter 5   WHEN THE CHIEF BECOMES THE FORCE: TRANSFORMATIONAL THEORY   50
Overview of the Theory   50
Importance of Charisma in Leaders   51

Meta-Analytic Studies on Transformational Leadership   53

Transformational Leadership in Military Contexts   53

Transformational and Transactional Leadership Styles and the Cases
of Chief Reuben Greenberg and Chief Daryl Gates   55

Chapter 6   PARAMETERS FOR EMPOWERMENT AND TRUST: STYLE
THEORY   66

Overview of the Theory   66

Ohio State University Leadership Studies   66

Michigan University Leadership Studies   66

Harvard University Studies   68

The Managerial Grid   68

Style as a Universal Theory   69

Style Theory and Military Leadership   69

Style Theory and Police Leadership   70

Style Theory and the Cases of Chief Earl Sanders and Chief Dennis
Nowicki   71

Chapter 7   WHEN THE EVENT IS JUST TOO MUCH TO HANDLE:
SITUATIONAL LEADERSHIP THEORY   81

Overview of the Theory   81

Situational Leadership Theory and Employee Burnout   83

Simplified Situational Leadership Theory   83

Politics and Situational Leadership Theory   83

Policing and Situational Leadership Theory   84

Situational Leadership Theory and the Cases of
Chief Charles A. Moose and Chief Bernard Parks   86

Chapter 8   DOING THINGS RIGHT OR DOING THE RIGHT THING:
CONTINGENCY THEORY   97

Overview of the Theory   97

Military Leaders and Contingency Theory   99

Cognitive Resources Theory   100

Leader Substitutes Theory   100

Vroom–Yetton Contingency Model   100

Dyadic Approach to Contingency Theory   101

Criticisms of Contingency Theory   101

Contingency Theory of Leadership and the Cases of Chief William O'Brien
and Chief Tom Koby   101

Chapter 9   WINNING MINDS AND HEARTS: PATH-GOAL THEORY   112

Overview of the Theory   112

Path-Goal Theory and Military Leadership   114

Path-Goal Theory and the Cases of Commissioner William Bratton and Chief Richard Pennington   114

Chapter 10   LEADERSHIP AND COMMAND OF THE CRITICAL INCIDENT: PSYCHODYNAMIC APPROACH   129

Overview of the Approach   129

Birth Order   131

Transactional Analysis   131

Psychohistory   132

Personality and Leadership   132

Psychodynamic Approach to Leadership and the Case of Commissioner Bernard Kerik   133

Chapter 11   SOLICITING AND ENTERTAINING 100 IDEAS: SKILLS APPROACH   141

Overview of the Approach   141

Learned Leadership Skills   141

Leadership in Context   142

Social Judgment as Social Intelligence   143

Emotional Intelligence   143

U.S. Army Studies   144

Skills Approach to Leadership and the Case of Commissioner Paul F. Evans   144

Chapter 12   CAREER OF LEADERSHIP: TRAIT THEORY   151

Overview of the Theory   151

Modern Applications of Trait Theory   152

Social Scientific Evaluation of Leadership Traits   154

Trait Theory and the Case of Chief George Edwards   156

Chapter 13   INTO THE FUTURE: THE BIG HAIRY AUDACIOUS GOAL AND CATALYTIC MECHANISMS   163

Anticharismatic Leadership   164

Level 5 Leadership   165

Choice of the Right People   167

Hedgehog Rule of Management   168

Big Hairy Audacious Goals (BHAGs) and the Formal and Informal Goals of Policing   168

Female Police Leaders—Some Thoughts for the Future   171

Chapter 14 IN THEIR OWN WORDS 175

Chief John DeCarlo 175

Deputy Chief William Fraher 176

Chief James R. Davis 177

Chief James R. Davis 179

Chief James R. Davis 181

Chief Robert F. Vodde 184

Chief Vincent Del Castillo 185

Inspector Walter Signorelli 187

Index 189

PREFACE

FIRST EDITION

This book is a product of years of research, teaching, and thinking about the importance of developing police leaders. No other profession has a more dire need for these skills. Leadership is, after all, a set of skills and traits that could and must be absorbed and learned by each police officer. The word "leadership" connotes many associations, and over the years and through many cultures, it has gained various definitions and interpretations.

Most of us think of leadership as something ultimately positive and very much desirable. However, leadership in general, and (in the context of this book) in police environments, can be also negative in nature. A Pakistani high-ranking police officer whom I met a few years ago, and whom I invited to give a guest lecture in my graduate class titled Police Leadership, started his comparative lecture about leadership in the Pakistani police force by referring to an example of his supervisor who, according to him, was a *good but corrupt leader*. The students and I looked at him in disbelief, trying to figure out if it was an improper use of the English language or if he really meant something that we had previously not discussed in class (nor something I had ever come across in a vast overview of literature on leadership). We asked for an explanation and received a very profound and telling one, especially with regard to the idea of effective policing. The good and corrupt police leader happened to be somebody who did things right—a true cold-blooded bureaucrat who did what was right for the police organization but did not do the right thing for the police officers.

I inquired during the guest lecture whether the word "corrupt" was too strong when used in reference to his actions. After all, to do things right for an organization sometimes entails putting the goal of the organization ahead of some personal needs of its employees. However, this was not the case. There is a tension between police organizations' formal goal of protecting the public and the informal goal of protecting politicians in power and maintaining the political status quo. This police supervisor was willing to engage in the most unethical behaviors in order to sustain the informal goal of police organizations.

The tensions between the formal and informal goals are an eternal conflict for police leaders. The Pakistani police leader took it a step further by identifying completely with the task and totally sacrificing the relationship aspect of the leader-follower exchange. This sacrifice earned him the label of good but corrupt leader. It was definitely a lesson I learned about police leadership.

A few years later, I taught the same class for New York Police Department officers. When one of the students complained bitterly about his supervisor, I asked him if he would consider his supervisor to be a good but corrupt leader. After some initial consternation on the part of the students and my in-depth explanation of the concept, he agreed—his supervisor was indeed a good leader but also a corrupt one.

A word of caution is mandatory here. In the same way that we accept many definitions for the word "leadership," there are also many definitions for the word "corruption." Corruption has many facets and angles, and not all of them have to do with being on the take or shaking the drug dealers or accepting a bribe from a speeding motorist. A corrupt leader can be very effective from the perspective of the informal goals of a police organization but at the same time remain corrupt from the perspective of human relations, whether within or outside the police organization.

This book highlights various leadership theories that were compiled by many authors who published in this area. Chief among them was a book by Peter Northouse, *Leadership: Theory and Practice* (2004). His book offers the most comprehensive overview of leadership theories applicable for police environments (although it was not written for that specific audience). Borrowing from his medley of leadership theories, I have customized the concepts for the police environment, matching these theories with police chiefs and commissioners to highlight the application of a given approach to the reality of everyday police life. Each chief and commissioner has a unique history, as do the events that make up each biography; so do the daily struggles of police supervisors and police officers, from the first day on the job. While the magnitude of events such as 9/11 is certainly not replicated in other environments, the potential for terrorist action is a definite problem for police leaders. The same can be said about events of police misconduct, problems with high crime rates, high-profile murder cases, racial tensions, and major innovations. Each police chief or commissioner has a unique story, but the leadership approach exhibited, whether conscious or subconscious, could be applicable in similar situations that are evolving and will continue to evolve on a daily basis in police work around the country.

In one of my previous works, a book on police training, I defined police leadership as the ability of each police officer, starting with the first day on the job, to take control of a situation on the street. Any type of situation that requires assertion of control would fall under this definition. As police officers progress in their careers, the level and degree of control change. Due to the fact that police work is based on the mandate to use coercive force to achieve compliance, police leadership is about the ability to take control. This book attempts to provide a template for police leaders, from street-level officers to the highest-ranking police chiefs, on how to look at a given situation, adopt an informed perspective—even when all that is available is the proverbial split second—and make the right leadership decision.

The book starts with an overview of police integrity and ethics and how they relate to the issue of leadership. Usually books on leadership end with the ethics chapter. For me, this is a fundamental mistake with regard to the entire concept of leadership. Leadership is about integrity and ethics, and this is where one should start the discussion of leadership concepts, not where it should end. Police leadership will frequently be immersed in a clash between the concepts of doing things right and doing the right thing, and the final judgment about what is the right approach is not something that this book aims to emphasize. Instead it does focus on the issue of situational integrity that is a product of a specific leadership approach of an individual chief—one that can be either emulated or modified by the readers, as customized for the relevant environment and situation.

The book ends with an optimistic approach—identifying the concept of the Big Hairy Audacious Goal of encouraging leaders to take hold of what seems like an impossible goal and to make it a reality. An approach that can manage the seemingly impossible is needed and appropriate in the police context, where effective policing is constantly challenged by the competing formal and informal goals of policing. If there ever was a Big Hairy Audacious Goal, bridging the gap between these two seemingly opposing goals would be it. Nevertheless, it is a goal worth achieving, and we owe it to the idea of democracy for which we all ultimately strive.

SECOND EDITION

This new edition brings many updates about the lives of the chiefs and the career moves that reflect, in part, the struggles and challenges encountered by them during their life on the job. In addition, some developments in thinking about police leadership issues are highlighted in chapter 1. The final chapter of the first edition, chapter 13, gained some volume, both in literal and metaphoric sense, by presenting 3 cases of female leadership, an aspect that was previously left out. While writing the first edition I struggled with the concept of a separate chapter dedicated to female leadership and felt that it is more important to concentrate on certain leadership situations that could be replicated rather than on the gender of the leaders. However, in hindsight, I felt that female leadership needs to be highlighted, even if the events depicted are not necessarily of the same high profile nature as the ones encountered by the males. Finally, one new chapter, chapter 14, depicts various leadership challenges presented by a number of police leaders, accounted for in their own words.

ACKNOWLEDGMENTS

As I mentioned in the first edition of this book, people rarely read the Acknowledgment pages, but the truth is that these pages are sometimes as important as the text itself, particularly for authors. Many things have changed in my life since the first edition went to press; therefore, I would like to modify this page and thank, first and foremost, Eric Krassow, Senior Acquisitions Editor for his vision for this book, and Tiffany Bitzel and Rex Davidson for their wonderful production support. In addition, my graduate research assistant, Brandon Roberts, contributed much to this edition with his diligent and insightful research and analysis, for which I am very grateful.

On a personal level, as always, my deepest gratitude goes to my daughters, Nellie and Mia, who inspire me with their love and support, each day anew. My parents, Dr. Lucja Sadykiewicz and Colonel (ret.) Michael Sadykiewicz continue to send e-mails of being "proud and happy" each time I inform them about my new publication and for this I would like to thank them as well. Finally, to my friends and police leaders around the world—thank you for being who you are and as such enabling me to be who I am.

Introduction
The Pentagon of Police Leadership

Many law enforcement agencies face real problems in identifying good leaders when a position becomes available. One reason for poor pools of candidates is because the development and training of leadership skills start too late. Another reason is because the potential leaders think more about management than leadership. Developing a leader is a long-term process in which encouragement, modeling, and support are combined with natural skills, and "the training should begin on the first day of an officer's career" (Bergner, 1998, p. 16).

This author firmly believes that leaders are both born and made. Since we have no control over the limited number of "born" leaders, our duty in establishing effective community-oriented law enforcement is to "make" more leaders. Baker (2000) suggests that the quality of police leadership directly affects the quality of life of police officers and the way they deliver their services. Police leaders have the potential to inspire officers but also to sabotage their efforts.

THE CASE FOR EARLY CAREER LEADERSHIP TRAINING

A common assumption is that police leaders are by definition the upper ranks. According to Swanson et al. (1998), police departments can be divided into three supervisory planes, each associated with a mix of leadership skills. These planes are based on ranks and consist of three basic skills: human relations, conceptual, and technical skills. Top management, composed of chiefs, deputy chiefs, and majors, predominantly use the conceptual skills, with marginal use of human relations and technical skills. Middle management, composed of captains and lieutenants, use many conceptual skills but also use human relations and technical skills. Finally, first-line supervisors, that is, sergeants, make relatively little use of conceptual skills, instead concentrating on the equal mix of human relations and technical skills (Swanson et al., 1998).

Criticisms of this depiction are numerous. For one, the degree to which each of the mix of skills is distributed among the three may vary, based on factors such as the size of an agency, its political environment, crime problems, and economic conditions. However, a much more important missing link relates directly to what Bergner (1998) mentions in his writings: Developing police leadership cannot ignore the leadership skills that line officers use on a daily basis. Line officers are the true leaders on the streets, using their leadership skills in daily encounters with the community, and police executives and policy makers need to realize it.

While human relations and technical skills play a very important role in line officers' work, conceptual skills, which are fundamental for problem solving, have particular applicability in the era of community-oriented policing. Many decision-making and other responsibilities previously left to middle and top management now have been placed on the shoulders of line officers; of course, not all such responsibilities were channeled "down." However, some common features of community-oriented policing—for example, decentralizing the chain of command, empowering employees, problem solving, and downsizing and flattening the organization—clearly require leadership skills.

Swanson et al. (1998) define conceptual skills as the ability to understand and interrelate various parcels of information, which are often fuzzy and appear to be unrelated. According to the authors, the standards for handling the information become less certain and the level of abstraction necessary to handle the parcels becomes greater as one moves upward through the chain of command. This author begs to differ with their interpretation. Line officers in the 21st century are expected to create innovative solutions, involve themselves in conflict resolution, solve problems of the most intricate nature—all of which require well-developed conceptual skills.

Furthermore, the reactive nature of police work is much more pronounced on the street than behind a desk. Top and middle management have the luxury of having to make the right or wrong decision in the comfort of their offices; line officers are frequently forced to make split-second decisions in front of a particularly watchful audience—the community they serve. Sometimes such decisions are literally a matter of life or death. More commonly, police work is frequently composed of split-second decisions when a police officer must take control of a situation where a wrong approach may escalate a potentially benign circumstance to irreversible damage. The definition of police leadership must include *the ability to make a split-second decision and take control of a potentially high-voltage situation that evolves on the street*. It is time to recognize this reality and equip officers with the necessary tools.

Vinzant and Crothers (1998) examine the leadership skills of public service workers—police and social workers—"who serve at the relative *bottoms* of their organizations and physically deliver services to the public" (p. 5). They also examine some of the prevailing leadership theories—trait, skill, situational, and transformational—as practiced in the scenarios that such employees meet in day-to-day work. Public servants' leadership can be divided into categories based on what kind of discretion the individuals have. They exercise leadership when they can exercise discretion in the *process* of responding to the incident (e.g., how to execute a search warrant), discretion over the *outcome* (e.g., whether to give someone a breathalyzer test when making a traffic stop), or both (e.g., how to respond to a domestic violence call; Vinzant & Crothers, 1998). In exercising process-oriented discretion, situational leadership skills are most relevant; in exercising outcome-oriented discretion, street-level leaders should use transformational models while considering the higher goal of society's greater interest when making their decisions on how to act.

If, indeed, leadership skills have become so profoundly important for line officers, one would expect to find an extensive leadership training module included in basic training at police academies. Unfortunately, this is not the case. Most basic training programs at police academies not only do not offer an extensive leadership module but also do not include even the most basic introduction to the leadership skills during basic training.

THE TRAINING GAP

The literature on leadership, in various contexts and forms, can fill the shelves of many book-cases; however, there is an absolute dearth in the area of leadership training and leadership theories that are applicable for and within police environments. Trainers and educators often integrate various ideas into their leadership training module, but the abundance of literature in this field can be overwhelming. The numerous leadership theories can be classified, in general, into four categories (Northouse, 2004):

1. "Great Man" and genetic theories
2. Traits approach
3. Behavioral explanations
4. Situational theories

Without clear examples from the world of policing, even overviews of the theories that appear to be most relevant for police environments will be of minimal utility for improving leadership in day-to-day policing. The traditional approach to what can be defined as "leadership-related issues" in police organizations is replete with terms relating more to management than to the pure issue of leadership.

An earlier book by this author, *Critical Issues in Police Training* (2002), discusses the need for leadership training to be introduced from the very first day of police basic training. This critical aspect of police basic training has been completely ignored by police organizations in the United States, as well as by those of many other countries that claim to have fully functional, democratic police forces. *Critical Issues* identifies some theoretical concepts of leadership and describes, in an overview, some of the rare existing training modules in the hope of inspiring much-needed programs that would be incorporated into basic academy training. However, *Critical Issues* only describes the contours of the problem. This book takes the issue a few steps further by identifying, in a much broader way, a number of leadership theories that are and should be applicable to various situations faced by law enforcement personnel on a daily basis.

Many police departments around the country experiment with a variety of leadership training modules, some developed and customized by the West Point Military Academy, others by the FBI National Academy, and still others endorsed or developed by the International Association of Chiefs of Police. Some departments opt to expose their supervisors to leadership training developed and offered by private consultants or academic institutions, such as colleges and universities with leadership institutes. Others attempt to customize the available programs and ideas to the needs of the specific environment. The NYPD, LAPD, Santa Barbara PD, Boston PD, a number of police departments in Washington, D.C., and Virginia (Arlington and Alexandria), and other predominantly large organizations lead these efforts.

Despite being clearly dedicated to the development of leadership skills, all such efforts continue to ignore the most needy target population—line officers. In their approach to leadership training, many organizations continue to utilize the traditional skills approach, based on a reactive instead of proactive training paradigm. (Reactive training focuses on delivering training, in various degrees of intensity, to the supervisory-level employees. A proactive training paradigm, advocated by this author, is leadership or any other new skills training delivered, in the most extensive manner, to line officers during their basic academy training.)

This proactive training paradigm is particularly applicable given potential flaws in the civil service system, which is usually used to select police supervisors. Proactive training can serve as a natural screening device and can identify early in their careers those individuals who have greater leadership potential. Every agency struggles with political interference and pressure from both inside and outside the organization when making appointments. While proactive training offers no solution to the problems that arise from the political nature of policing and the influence of politics on police leadership, providing leadership training to each and every member of the organization at the very outset of their careers will produce a pool of potential candidates that will be hard to ignore, even by the savviest of politicians.

In the classroom, this author has frequently come across the best and the brightest students who, due to the bureaucratic nature of their police organizations, remained line officers for years before being given the opportunity for advancement. If leadership skills, like any other type of skill, are better absorbed by some than by others, police organizations will benefit by identifying those people early and taking advantage of their talents.

The focus of this book is as much on the theories as it is on their practical implementation. It categorizes and describes many of the applicable relevant theories and then illustrates those theories with practical examples from U.S. police departments. The examples focus on the highest echelon of the police hierarchy—the police chiefs and commissioners. The question that begs to be answered is, why does a book on police leadership that so strongly emphasizes the need for early police leadership training focus on events and situations in which the major players are the top executives and not the line officers?

The answer is pretty simple. It is very hard to learn from mundane examples that lack prominence, mystery, and real drama as experienced by others. Yes, this author could have included scenarios or even real-life stories from the daily routine encounters of line officers on the job: the arrival at a crime scene, an angry demonstrator, a traffic violation, a high-speed pursuit, or a domestic violence encounter. The definition of police leadership is *the ability to make a split-second decision and take control of a potentially high-voltage situation that evolves on the street*. Such situations have to be absorbed by the recruits from the standpoint of what is expected of them as a leader. A recruit, entering the police academy, will not absorb easily or quickly the notion of being a leader on the first days on the job. It will be too amorphous a concept, and the focus will be more on understanding practical operating procedures.

However, if one teaches leadership through the prism of established position, then it will be much easier for the recruit to understand the nature of the problem, grasp the relevance of abstract concepts, and apply them to the lower tiers. For example, the struggles with issues of integrity and ethics faced by the police chief of New Orleans can be easily understood and analyzed by a police recruit in the same way that they can be easily understood and analyzed by any student of police science, whether or not he/she is a police officer. On the other hand, if one were to start learning about the problems of integrity and ethics from the street-level perspective—for example, by looking at a line officer being offered a bribe by a motorist—it becomes too personal, close, scary, and even immobilizing because it requires an almost immediate split-second response with regard to what is right and wrong, without having the room not only to understand and analyze but also to be able to argue the pros and cons of doing things right versus doing the right thing. When the examples or scenarios one is supposed to learn from hit too close to home, our reactions tend to lean more toward what we know is expected of us rather than toward what we really feel and think about the issue at hand. That is why it will be much easier, for example, to look at the behavior of Chief O'Brien (who decided not to involve the mayor of Miami in his operational decision regarding Elian Gonzales) and to analyze the pros and cons of his leadership approach in this given situation than to learn the same lessons from a situation that involves a line officer having to make a decision that goes against the departmental rules but that makes moral or ethical sense to the line officer. Learning leadership concepts from top-ranking supervisors, and then adapting them to line-level situations, is a much safer approach to this complex topic.

There is one disclaimer that cannot be repeated often enough. Anyone's career is a complex three-dimensional entity, and no matter how many pages are dedicated to describing it, they will not capture every phase and nuance of it. The cases that were chosen to illustrate the theories depict only a small facet of any chief's career—isolated situations or particular characteristics. Furthermore, no one real-life situation will ever exactly model the ideal of an abstract theory. The point of applying the theories to the chiefs is to open discussion, to further explore the theories through practical examples. There is nothing written in stone, and disagreement and debate will only further deepen understanding.

THE PENTAGON OF POLICE LEADERSHIP

There are five model components leading to effective leadership, introduced here as the *pentagon of police leadership,* that can be emulated in police training academies. The five components of the pentagon are recruitment, selection, training, supervision, and discipline (see Figure 1–1).

In the ideal world of police profession, the Pentagon of Police Leadership would be composed of equal-length prongs, where each prong represents a substantial, and equally resource loaded, approach to maintaining a professional organization. However, in the case of political pressure to staff the academy class or to provide more officers on the street to give the public an impression that officers are present on the streets, the first two prongs are frequently short-handed, and the prongs of supervision and discipline are extended (turning the supervision to more oppressive and the discipline to more severe); only a proper allocation to the prong representing training can mitigate the decline of a model democratic police organization. Turning the regular pentagon, in which all the prongs are even, to an irregular one where one or more prongs require more attention, and, in police organizations, this could translate into a more oppressive

The Pentagon of Police
Leadership: A Regular Pentagon

Recruitment   Selection

Discipline   Training

Supervision

FIGURE 1-1 Pentagon of Police Leadership

Source: J. Bjorken, 2004, based on the model by M. H. Haberfeld.

work environment, is not something that law enforcement organization should strive for. However, given the traditional approach to training within police organizations, where under a financial duress the first cuts are almost always directed at the training resources, it is imperative to find another way to maintain the level of training that is a necessary component of an effective and democratic policing.

Meta-Leadership Approach

Such a solution can be possibly implemented by adopting the concept of "Meta-Leadership" that was originally developed by Marcus, Ashkenazi, Dorn, and Henderson (2009). The main concepts of Meta-Leadership can be summarized in the following points:

- *Meta-leadership* is a term used to refer to a style of leadership that challenges individuals to think and act cooperatively across organizations and sectors. Meta-leaders develop ways to engage in interactions outside the scope of their traditional professional boundaries, providing inspiration, guidance and momentum for a course of action that spans organizational lines.
- *Meta-leadership* became a vital ingredient for effective emergency preparedness and response. In a crisis, business, government, and nonprofit leaders are thrown together, exchanging information, directing resources and managing systems and personnel. The commitment of leaders across sectors to contribute to and guide a coordinated strategy can be a critical factor in the overall success of the response.
- *Meta-leadership* identifies the process and practice of leaders based on three functional components: 1) A comprehensive organizing reference to understand and integrate the many facets of leadership; 2) a strategy to engage collaborative activity; and 3) a cause and purpose to improve community functioning and performance. Leonard J. Marcus, Ph.D.; Isaac Ashkenazi, M.D., M.P.A.; Barry C. Dorn, M.D., M.H.C.M.; and Joseph M. Henderson (http://meta-leadershipsummit.org/Files/meta/2011/FiveDimensionsofMeta-LeadershipComprehensiveBrief.pdf)

Adopting the meta-leadership solutions to leadership training would allow for integration of many actors into the process, starting with the line officers who need to be socialized to the

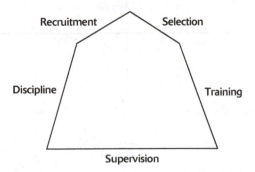

FIGURE 1-2   An Irregular Pentagon

needs of professional development that is not always initiated by the organization, followed by the responsibility, on the part of the supervisors, to reach out to external bodies and engage them in the delivery of the ongoing leadership development. (This author elaborated more on this point in her article, "Doing More with Less—Solutions to Budget Cuts to Municipal Police Departments Based on the Meta-Leadership Concept," published by the Austrian Police Journal, SIAK, in 2011).

The efforts to develop future police leaders can certainly be traced to the early stages of recruitment and selection. The process of *recruitment* and *selection* of recruits with potential to lead, on the streets as well as behind desks, is just the initial stage of the five-prong approach to be applied to the development of the police leadership cadre. By offering the leadership *training* (the third prong) at the basic academy level, it will be possible to identify the level of readiness of the officers for future supervisory positions within the department and to expose them to advanced leadership training, customized to their level of readiness and individual style, further down the road. Advanced leadership training will then concentrate *primarily on the issues pertinent to the functions of supervision and discipline,* the fourth and fifth prongs of the pentagon of police leadership.

This book is a tool for that process. By evaluating the ways in which police recruits look at the cases presented in this book, the ways they analyze behaviors and look for other and/or better leadership approaches, the academy trainers will be able to distinguish those who have the highest potential to climb the supervisory ladder and in doing so identify, at the earliest stage, those who should be given additional opportunities to hone their natural predispositions.

This book discusses some of the prominent theories of leadership that are applicable to the policing context, in the format of case studies accompanied by an in-depth overview of the theories. The basic models put forward by leading academics and other theorists are summarized. Then the biography of a police chief is used to illustrate how that particular theory can be applied to practical circumstances. The chapters end with an invitation to consider how other theories the student has learned might be applied to the chief's or commissioner's circumstance.

Police work is full of temptations to err and to experience the consequences of those errors of judgment on the most profound level, consequences that touch not only other people's lives but also the officers'. The police chiefs and commissioners who were chosen to illustrate the theories were probably not even conscious of the fact that their decision-making process was informed by one leadership approach or another. Some of them, as well as some readers, will disagree with the label attached to them by the author. The true beauty of social science is the ability to argue and dispute many notions and perceptions. Experimentation with concepts and ideas is truly a building process in this complex maze of skills, traits, and approaches that can and will make the life of police officers and those they police much more worth living.

The reader of this book, whether a police recruit, trainer, ranking officer, or student with interest in police work, is strongly encouraged to familiarize himself/herself with the following summary of the basic tenets of each leadership theory (see Table 1-1). This will serve as a foreshadowing and will also be useful when reading the case studies. It is important for the reader to be able to grasp the basic differences among the theories in order to fully appreciate all the subtleties and nuances that differentiate the approaches. This preparation will enable the reader to go back and forth and to experiment with a number of alternative approaches that could/would be beneficial in a given situation for a specific police chief or commissioner and from there to be able to deconstruct and translate the example into a street-level reality, one that each one of us can and will face on a daily basis.

TABLE 1-1	Leadership Theories
Theory	**Basic Concepts**
Team	• Leadership involves the solving of complex organization problems. • Leadership of teams involves team performance and team development. • Leaders must monitor and diagnose problems that could impede organization and group performance, plan appropriate solutions, and then implement them, in an effort to adapt to changing conditions (Zaccaro et al., 2001). • Four processes of leadership determine team productivity: cognitive, motivational, affective, and coordination (Zaccaro et al., 2001). • Self-managed teams involve either rotated leadership and democratic decision making or emergent leadership in which an individual steps forward to assume the leadership role (Erez et al., 2002).
Leader–member exchange	• This theory focuses on the interactions between a leader and subordinates. • Each subordinate has a dyadic relationship of vertical linkage with the leader (Liden & Graen, 1980). There are two types of relationships, with subordinates forming an in-group and out-group in relation to the leader (Graen & Uhl-Bien, 1991). • Subordinates in the in-group go beyond their mere job descriptions and as a result receive more information, have better confidence, and elicit more concern from the leader (Graen & Cashman, 1975). • Effective leaders should develop in-group exchanges with all subordinates if possible (Graen & Uhl-Bien, 1991).
Transformational	• Transformational leaders raise motivation and morale in subordinates (Burns, 1978), often through charisma (House, 1976). • Transformational leaders cause followers to engage beyond mere job description by raising consciousness, motivating a transcendence of self-interest, and promoting the attainment of higher-level needs (Bass, 1985). • As social architects, transformational leaders shape shared meanings in organizations and direct mutual values and norms (Bennis & Nanus, 1985).
Transactional	• Transactional leaders exchange things of value (e.g., contingent reward: work for a paycheck) with subordinates. Both leaders and subordinates engage in the transaction for their own self-interest (Kuhnert, 1994). • Management-by-exception characterizes the leadership style. Active management-by-exception refers to daily monitoring by the leader of problems with subordinate performance. Passive management-by-exception involves intervening only when standards are in jeopardy (Northouse, 2004).
Style	• Leadership effectiveness is dependent on behavior. • There are two broad categories of behavior: task and relationship (Northouse, 2004), for example, consideration and initiation of structure (Stogdill, 1974) or concern for structure and concern for people (Blake & Moulton, 1985). • According to Blake and Moulton (1985), the intersection of the two behavior dimensions forms a grid of styles of management. • Each leader has one style that is used in all situations.

Theory	Basic Concepts
Situational	• The type of leadership behavior that is most appropriate depends on the situation. • To determine the correct behavior, the leader takes into consideration the competency and commitment of subordinates. • According to Hersey and Blanchard (1977), there are four main leadership styles: delegating (low directive-low supportive), supporting (low directive-high supportive), coaching (high directive-high supportive), and directing (high directive-low supportive). • A leader may have to use a variety of styles throughout a career, depending on situations faced.
Contingency	• Fiedler's (1967) contingency theory attempts to match leadership styles to specific situations. • There are three variables associated with a situation: leader–member relations (confidence and loyalty), task structure (clarity of task), and position power (amount of leader authority). Certain leadership styles (task versus structure) are better in certain situations. • Therefore, not all leaders are good for all situations. Leadership is effective to the extent that the leader is compatible with the situation (Northouse, 2004).
Vroom-Yetton Model	• The Vroom and Yetton (1973) contingency model focuses on the autocracy-participation dimension of leadership style. The model looks at leadership as a problem to be solved. In order to handle the problem, the leaders may or may not share their decision making with subordinates. • The level of participation represents a continuum of five levels of employee participation, from no participation to democratic participation. The leader decides which of the five levels to adopt based on seven rules, which are designed to either protect the quality of the decision or protect the acceptance of the decision by subordinates (Vroom & Yetton, 1973). • According to the model, a single leader should be flexible enough to vary the style across the continuum as needed; therefore, a person has the capacity to be both autocratic and participative (Bryman, 1986).
Leader Substitutes	• This theory is based on the notion that group processes can produce effects similar to interpersonal, hierarchical leadership (Tosi & Kiker, 1997). The idea was developed by Kerr and Jermier (1978) from the observation that situational factors can actually negate a leader's ability to be effective. • The model makes a distinction between two kinds of situational variables: substitutes (what makes leader behavior unnecessary or redundant) and neutralizers (what nullifies the effects of the leaders) (Yukl, 1994). • Formal leadership may be unnecessary in the presence of capable subordinates (Kerr & Jermier, 1978).
Path-goal	• Path-goal matches leadership style with the characteristics of subordinates and the work environment. • Subordinates are only motivated when they believe they can perform their work, that they can achieve their goals, and that there are benefits to doing so. The leader's task, therefore, is to remove obstacles to success, to coach subordinates, and to instill motivation (House, 1971).

(continued)

TABLE 1-1	Leadership Theories (*continued*)
Theory	**Basic Concepts**
Psychodynamic	• This approach was derived from Freud (1921, 1939). Deep-seated emotional factors influence leaders. • Childhood family experiences affect leaders who always take a paternal role with subordinates. Based on family experiences, leaders exhibit behaviors along continuums of authoritative to permissive, of supportive to critical, of dependent to counterdependent to independent (Northouse, 2004). • Based on Jung (1923), leaders must understand their own shadow self, or their unconscious, including repressed needs and desires, in order to effectively lead. • Berne (1961) posited that an effective leader operates out of an adult ego state, reflecting emotional maturity. • Because of the different psychodynamic outcomes leaders embody, leaders will match better with some subordinates than others, depending on subordinates' psychodynamic variables. Leaders' awareness of their own personality types will help them relate to subordinates more effectively (Kroeger & Theusen, 1992).
Skills	• A leader's success is determined by skills and abilities, which are not innate but which can be learned and developed. • Leaders must be able to use their skills to solve complex organizational problems. • Technical, human relations, and conceptual skills are the primary skills, according to Sheriff (1968). • In addition, social intelligence (Kobe et al., 2001) and emotional intelligence (Feldman, 1999) are leadership skills that produce effectiveness.
Trait	• Individuals' traits determine their success as leaders. Traits are innate, according to the "Great Man" theory, on which the trait theory is based. • Traits associated with leadership are intelligence (Lord et al., 1986; Mann, 1959; Stogdill, 1948), insight, responsibility, persistence, self-confidence, initiative, sociability (Stogdill, 1948), cognitive ability, motivation, drive, task knowledge, and integrity (Kirkpatrick & Locke, 1991).

References

Baker, T.E. (2000). *Effective police leadership: Beyond management*. New York: Looseleaf Law Publications, Inc.

Bass, B.M. (1985). *Leadership and performance beyond expectations*. New York: Free Press.

Bennis, W.G. & B. Nanus (1985). *Leaders: The strategies for taking charge*. New York: Harper & Row.

Bergner, L.L. (1998, November). Developing leaders begins at the beginning. *Police Chief*, 16–23.

Berne, E. (1961). *Transactional analysis in psychotherapy*. New York: Grove.

Blake, R.R. & J.S. Moulton (1985). *The managerial grid III*. Houston, TX: Gulf.

Bryman, A. (1986). *Leadership and organizations*. London: Routledge & Kegan Paul.

Burns, J.M. (1978). *Leadership*. New York: Harper & Row.

Erez, A., J.A. Lepine & H. Elms (2002). Effects of rotated leadership and peer evaluation of the functioning and effectiveness of self-managed teams: A quasi-experiment. *Personnel Psychology*, 55, 929–948.

Feldman, D.A. (1999). *The handbook of emotionally intelligent leadership: Inspiring others to achieve results*. Falls Church, VA: Leadership Performance Solutions Press.

Fiedler, F.E. (1967). *A theory of leadership effectiveness*. New York: McGraw-Hill.

Freud, S. (1921; 1959). *Group psychology and the analysis of the ego*. London: Hogarth Press.

_____ (1939; 1964). *Moses and monotheism*. London: Hogarth Press.

Graen, G. & J.F. Cashman (1975). A role making model of leadership in formal organizations: A developmental approach. In J.L. Hunt & L.L. Larson (Eds.), *Leadership frontiers* (pp. 143–165). Kent, OH: Kent State University Press.

_____ & M. Uhl-Bien (1991). The transformation of professionals into self-managing and partially self-designing contributions. Toward a theory of leader-making. *Journal of Management Systems*, 3 (3), 33–48.

Haberfeld, M.R. (2002). *Critical issues in police training*. Upper Saddle River, NJ: Prentice Hall.

Hersey, P. & K.H. Blanchard (1977). *Management of organizational behavior: Utilizing human resources*. Englewood Cliffs, NJ: Prentice Hall.

House, R.J. (1971). A path-goal theory of leader effectiveness: Lessons, legacy and a reformulated theory. *The Leadership Quarterly*, 7 (3), 323–352.

_____ (1976). A 1976 theory of charismatic leadership. In J.G. Hunt & L.L. Larson (Eds.), *Leadership: The cutting edge* (pp. 189–207). Carbondale: Southern Illinois University Press.

Jung, C. (1923). *Psychological types*. New York: Harcourt & Brace.

Kerr S. & J. Jermier (1978). Substitutes for leadership: Their meaning and measurement. *Organizational Behavior and Human Performance*, 22, 374–403.

Kirkpatrick, S.A. & E.A. Locke (1991). Leadership: Do traits matter? *The Executive*, 5, 48–60.

Kobe, L.M., R. Reiter-Palmon & J.D. Rickers (2001). Self-reported leadership experiences in relation to inventoried social and emotional intelligence. *Current Psychology*, 20 (2), 154–163.

Kroeger, O. & J.M. Theusen (1992). *Type talk at work*. New York: Delacorte.

Kuhnert, K.W. (1994). Transforming leadership: Developing people through delegation. In B.M. Bass & B.J. Avolio (Eds.), *Improving organizational effectiveness through transformational leadership* (pp. 10–25). Thousand Oaks, CA: Sage.

Liden, R.C. & G. Graen (1980). Generalizability of the vertical dyad linkage model of leadership. *Academy of Management Journal*, 23 (3), 451–465.

Lord, R.G., C.L. Devader & G.M. Alliger (1986). A meta-analysis of the relationship between personality traits and leadership perceptions: An application of validity generalization procedures. *Journal of Applied Psychology*, 71, 402–410.

Mann, R.D. (1959). Review of the relationship between personality and performance in small groups. *Psychological Bulletin*, 56 (4), 241–270.

Marcus, L.J., I. Ashkenazi, B.C. Dorn & J.M. Henderson (2009). The five dimensions of meta-leadership: National preparedness and the five dimensions on meta-leadership. Retrieved from http://meta-leadershipsummit.org/Files/meta/2011/FiveDimensionsofMeta-LeadershipComprehensiveBrief.pdf.

Northouse, P.G. (2004). *Leadership: Theory and practice* (3rd ed.). Thousand Oaks, CA: Sage.

Sheriff, D.R. (1968). Leadership skills and executive development: Leadership mythology vs. six learnable skills. *Training and Development Journal*, 22 (4), 29–34.

Stogdill, R.M. (1948). Personal factors associated with leadership: A survey of the literature. *Journal of Psychology*, 25, 35–71.

_____ (1974). *Handbook of leadership: A survey of theory and research*. New York: Free Press.

Swanson, C.R., L. Territo & R.W. Taylor (1998). *Police administration: Structures, processes, and behavior* (4th ed.). Upper Saddle River, NJ: Prentice Hall.

Tosi, H.L. & S. Kiker (1997). Commentary on "substitutes for leadership." *Leadership Quarterly*, 8 (2), 109–112.

Vinzant, J.C. & L. Crothers (1998). *Street-level leadership: Discretion and legitimacy in front-line public service*. Washington, D.C.: Georgetown University Press.

Vroom, V.H. & P.W. Yetton (1973). *Leadership and decision-making*. Pittsburgh: University of Pittsburgh Press.

Yukl, G. (1994). *Leadership in organizations* (3rd ed.). Englewood Cliffs, NJ: Prentice Hall.

Zaccaro, S.J., A.L. Rittman & M.A. Marks (2001). Team leadership. *The Leadership Quarterly*, 12, 451–483.

Integrity, Ethics, and Police Leadership

Police integrity is defined in many different ways by many different actors. For the purpose of this chapter, the author will be using the following definition, developed during a study of 30 police departments in the United States that were surveyed by the author and her colleagues between the years 1996 and 1997:

> Police integrity is the normative inclination among police to resist temptations to abuse the rights and privileges of their occupation. (Klockars et al., 1997, 2002)

Using the word "integrity" appears to be less threatening for police organizations than the word "ethics." Ethics somehow connotes that if somebody is in need of ethical training or exposure to ethical decision-making processes, there is something deficient in that area that needs to be addressed and/or corrected. Therefore, substituting "integrity" for "ethics" creates a somewhat more appealing notion—that the ethical foundation is in place and that what we are really looking for is some form of enhancement and maintenance of otherwise ethical environments. "Integrity" is simply a much more user-friendly word than "ethics," and this is an especially poignant point with regard to police environments.

This chapter will provide a general overview of literature related to the concept of ethics within the leadership context and proceed to identify ways police leaders could and should create and maintain environments of integrity in their organizations. These practical applications are as relevant for first-line supervisors as they are for chiefs and commissioners; it is not a coincidence that this chapter opens the door to various theories of and approaches to leadership within police organizations.

Leadership scholars underemphasize leadership ethics. According to Ciulla (1995), "good leadership" implies two realms: 1) effective leadership and 2) ethical leadership. Yet a majority of leadership studies focus only on the first consideration. Ciulla argues that the focus on scientific assessments of leadership has neglected the softer analysis required of leadership ethics. "[N]o matter how much empirical information we get from the 'scientific' study of leadership, it will always be inadequate if we neglect the moral implications" (Ciulla, 1995, p. 14). Ciulla explains that each of the normative theories in leadership studies has inherent philosophical underpinnings that beg ethical questions.

Ethical considerations are important because confidence in leaders, and thus their effectiveness, falters if followers do not perceive a sense of ethical responsibility from leaders (Martinez-Carbonell,

2003). Maintaining trust in leadership is a critical component in a healthy organization. "If an atmosphere of misinformation and distrust exists, work activities may continue even though the workplace is hampered by rumors and suspicion. However, motivation and efficiency will be hindered" (Eckhart, 2003, p. 6).

ETHICS AND TRANSFORMATIONAL LEADERSHIP

Ethics as it relates to transformational leadership has been the subject of the bulk of articles on leadership ethics in general (see the table in Chapter 1 and Chapter 5 for an overview of transformational theory). Burns's (1978) *Leadership* focuses on the notion that people can be lifted up to become their better selves through the transforming power of leadership. Leaders are to encourage followers to develop a set of values that emphasizes justice, liberty, and equality. This is what gives transformational leadership its moral purpose. The moral calling of transformational leadership is then contrasted with transactional leadership. Burns finds transactional leadership to lack a moral center because it is inherently uncritical of the people involved in the leadership exchange. It is enough that people operate in their own self-interest; there is no encouragement to ascend to higher-order moral functioning. Giampetro-Meyer et al. (1998) come to a similar conclusion, arguing that transactional leadership can lead to ethically questionable behavior on the part of leaders. This is because of what they term "craft ethics" (p. 1731), in which leaders do not reflect on what they personally think is right but merely follow what those on the job or in the workplace believe is ethical.

Similarly, for Kuhnert and Lewis (1987), transformational leaders transcend self-interested goals to a more mature set of values, culminating in a sense of universal ethical principles of fairness, equality, and justice. These values form ends in themselves for the mature leaders. In fact, moral reasoning has been empirically connected to leadership effectiveness, using a measure of moral reasoning (Defining Issues Test [DIT]) and subordinates' ratings of leaders in a sample of managers. Moral reasoning is considered a cognitive function that can develop over a life span. Leaders with the cognitive capacity to reason morally will use this higher degree of transformational leadership and will be more likely to behave in a way that serves the collective good. Leaders who use the transactional style of leadership, however, do not become more effective regardless of whether they have the capacity to reason morally. This is because, by definition, transactional leadership is merely a dyadic exchange to satisfy individual needs (Turner et al., 2002). Contrastingly, however, in a study of U.S. military cadets, emergence of moral reasoning abilities in the first year of military training was not predictive of emergence as a leader in later years. This suggests that in strong hierarchical organizations, moral reasoning may not be critical to performing well as a leader. Such environments may be inherently transactional, lacking the flexibility for moral reasoning and transformational leadership to come into play (Atwater et al., 1999).

One of the problems with the above conception of the moral superiority of transformational leadership is that it assumes that leaders know and operationalize the values to be instilled in followers and that they have the will to proceed as such:

> An analysis of the ethics of transformational leadership is thus closely connected to the question of why leaders behave immorally. On the standard view in applied ethics and moral theory, ethical failures are essentially volitional, not cognitive. We behave unethically because of problems of will, not because of problems of belief and knowledge. (Price, 2003, p. 69)

Aside from problems of will, Price argues that leaders can behave immorally because they are blinded by their own values. Transformational leadership may actually lead to a dangerous moral superiority in which leaders believe they are justified in doing as they wish, even exempting themselves from ethical behavior. This is similar to Bass and Steidlmeier's (1999) notion of the pseudo-transformational leader who purports to be moral but in fact is manipulative and victimizing. Price explains that leadership is not so black and white and that the real test of authentic transformational leadership is whether leaders can apply moral requirements to themselves.

However, Kanungo (2001) argues that transformational and transactional leadership can both be ethically based. They reflect two different value systems and motives. Unlike most other theorists, Kanungo does not believe that transformational leadership is morally superior. For him, transformational leaders possess an organic worldview characteristic of a deontological perspective. Such a perspective asserts that leaders' actions have a morally intrinsic value. An act is moral when, as a means to an end, it is executed with a sense of duty to others and guided by Kantian pure reason; this is referred to as genuine or moral altruism:

> While acting out of a sense of duty, the leader is prepared to suffer the harmful consequences for him/her self. Sometimes, the leader knowingly causes harm to him/her self, a strategic move that convinces others of the leader's unbending commitment to the organizational objectives. (Kanungo, 2001, p. 261)

In contrast, transactional leaders reflect teleological ethics in which ends or outcomes are emphasized. The leader has an atomistic worldview, meaning that he/she feels separate from others, with an emphasis on independence rather than interdependence. Under this paradigm, actions that provide the most benefit to subordinates in the end are considered moral; this is referred to as mutual altruism. In this mode of influence, transactional leaders are only unethical if they follow their own self-interest without providing a beneficial end result for subordinates.

INTEGRATION OF BEHAVIOR AND VALUES

Other theorists have written about the problem of integrating behavior and values in the context of leadership in general rather than merely in transformational leadership. Weber (1989) believes that organizations can potentially stifle the moral maturity of leaders in favor of other organizational values such as competitiveness and efficiency. Although a leader may be able to reason in a morally mature fashion, research has not established a connection between this and the capacity or opportunity to behave morally.

Heifetz (1994) writes about leadership as the capacity to behave in a manner that is commensurate with moral reasoning and values. Leaders assist subordinates in resolving fundamental value conflicts that occur in organizations in which people have different job functions, social roles, and personal interests. In essence, leadership is adaptive work that "consists of the learning required to address conflicts in the values people stand for and the reality they face" (Heifetz, 1994, p. 22). Further, effective leadership involves fostering situations in which followers can learn to reconcile values with reality and thus develop their moral compasses.

STEWARDSHIP MODEL

Under the stewardship model, leaders serve their followers and put their needs above their own self-interest. In other words, the leader becomes a steward whose job is to protect the interests of subordinates:

> [T]he steward is motivated by deep intrinsic values based upon a moral theory that a prima facie priority of interests is owed to those who have a stake in the organization's success. The underlying mechanism is that the steward is driven by an underlying social contract. (Caldwell et al., 2002, p. 161)

This is similar to Greenleaf's (1977) servant leadership, which centers around the need for leaders to be attentive to the interests of their followers. Under this model, leaders are charged with nurturing subordinates, helping them to develop their intellect, independence, and personal leadership abilities. In essence, the leader is a servant to followers. Giampetro-Meyer et al. (1998) point out that Greenleaf's conception is particularly moral: "Servant leaders are concerned with the least privileged in society and strive to help others grow as persons. . . . Servant leaders show awareness of how the decisions they inspire affect others" (p. 1734).

Caldwell et al. (2002) explain that an effective steward is a "facilitating idealist" (p. 156). Stewards recognize that reality is complex and ultimately unpredictable. They also accept that individual perception is imperfect and limited even in the best of circumstances and with the most conscientious of stewards. However, leaders as stewards proceed to make decisions that serve the best interest of all stakeholders. This requires adaptability, flexibility, and commitment to continual learning.

Stewardship also requires a self-exploration of beliefs, values, and assumptions that inform stewards' decision-making process. Schein (1977) provides a framework for assessing these assumptions. One should reflect on these areas:

- Beliefs about the self and about the ways stewards view their own talents, goals, and worth
- Beliefs about others, including beliefs about the nature of people and organizations, and about stewards' social responsibility
- Beliefs about the past and how it has influenced current events
- Beliefs about the current reality as a process of filtering data and interpreting information
- Beliefs about the future, including goals and spiritual tenets

ETHICAL LEADERS' CHARACTERISTICS

Ethical leaders have been characterized in a variety of ways. The main characteristics fall into three broad categories: being nurturing and service-oriented, being just and honest, and being committed to common goals.

Nurturing and Service-Oriented

Burns (1978) indicates that leaders must instill in subordinates an awareness of their needs and values; they must be nurturing. Similarly, Greenleaf (1977) emphasizes the service of followers, encouraging them to develop their values and moral visions. Caldwell et al. (2002) emphasizes that effective leaders are committed to all stakeholders in a given issue and pursue integrative solutions. Further, leadership should not force followers to a certain end, nor should it ignore the interests of subordinates (Bass & Steidlmeier, 1999).

Just and Honest

Effective transformational leaders have a sense of social responsibility and fairness (Kanungo, 2001; Kuhnert & Lewis, 1987), follow established rules, and are objective evaluators of the facts underlying any decision (Caldwell et al., 2002). Lying is often cited as one of the most destructive behaviors that a leader can exhibit because it leads to a breakdown in trust, the foundation of any relationship. "One of the critical 'cures' for . . . an organization is truthfulness. . . . Being truthful involves taking active steps to promote understanding, especially when misperceptions persist" (Eckhart, 2003, p. 6).

Committed to Common Goals

Burns (1978) and Bass and Steidlmeier (1999) describe transformational leadership as the development of common goals that are beneficial to both leaders and followers. Moreover, a sense of the common good should be broadly interpreted by leaders and includes the public interest as well as the interests of leaders and their subordinates.

THE TRIANGLE OF POLICE INTEGRITY

In the policing environment, three general topics—recruitment, selection, and training—are viewed in the context of integrity and corruption. The very essence of this chapter has to do with and is predicated on the assumption that whatever does not lead to integrity will, by default, actually or potentially lead to corruption. The word "corruption" connotes numerous interpretations, some of which include noble cause corruption, abuse of power, excessive use of force, and corruption for gain (Crank and Caldero, 2000; Klockars et al., 2002; Punch, 1999). For the purpose of this chapter, however, corruption will be viewed as any misconduct or inaction exhibited by a police officer that can be directly tied to poor and/or inadequate mechanisms of recruitment, selection, and training.

There are three crucial topics in which the theoretical models of ethics in leadership apply to policing. These are recruitment, selection, and training—three of the five prongs of the pentagon of police leadership (discipline and supervision are the other two). Each has a relationship to integrity.

Recruitment

The relationship between recruitment and integrity can be approached from a number of theoretical angles. First, the overall endeavor devoted to the recruitment process indicates whether a given agency views the outreach effort as an important component of its internal infrastructure. An agency that invests significant resources, from qualified personnel to adequate facilities, in its recruitment process sends a powerful message about its professional standards to potential applicants.

The intensity and variety of outreach efforts are closely related to integrity. An agency that attempts to recruit in narrowly defined environments, such as military facilities or community colleges, or that narrows its recruitment effort to a recruitment office within that agency can be considered to have a much skewed interest in the quality of potential applicants. On the other hand, an agency that is highly engaged in proactive recruitment, attempting to reach out to applicants of diverse socioeconomic, educational, ethnic, racial, and other backgrounds, sends a signal about its ethical orientation and motivation (orientation and motivation which might

Framing the Problem

1. NOBLE CAUSE CORRUPTION (Crank and Caldero, 2000)
   a. ***Noble cause:*** Not directly making choices for personal gain
   b. ***Victim:*** Acting on behalf of victims and doing "the right thing"
   c. ***"Tower":*** Choosing to run toward the sniper on the tower instead of away from him and belief that this gives one the right to act in a certain way—not necessarily by the book
2. ADAPTATION (Crank, 2004)
   a. ***The asshole:*** Viewing everybody, including self, as such
   b. ***Cynicism:*** Resulting from viewing the worst that humans are capable of
   c. ***Means-end conflict:*** Doing the right thing or doing things right
3. ECONOMIC CORRUPTION (Kleinig, 1996)
   a. ***Stages—the slippery slope:*** Starting with something small that leads to more
   b. ***Typology:*** Using different types of corruption for gain
   c. ***Corruption of authority:*** Being involved in the following:
   Kickbacks
   Opportunistic thefts
   Shakedowns
   Protection of illegal activities
   Traffic fixes
   Misdemeanor fixes
   Felony fixes
   Direct criminal activities
   Internal payoffs

originate from political pressures but which, nevertheless, are omnipresent). A recruitment staff composed of selectively chosen career-oriented officers, as opposed to a small number of unmotivated employees, projects an image of professionalism and high standards. Recruitment offices located in temporary or poorly maintained facilities, equipped with old or broken furniture, cast an image of neglect on the force they represent. On the other hand, well-equipped, modern, and permanently located offices project a sense of importance and seriousness about the activities taking place inside.

In sum, a combination of resource allocation, marketing techniques, and careful planning, exemplified by the intensity, diversity, and quality of effort devoted to recruitment, bears an important correlation to the integrity of a department.

Selection

The actual and potential relationship between the selection process and integrity is the second theme. A number of theoretical questions need to be addressed in order to clarify the importance of the selection process. To begin with, if it is a fundamental premise that there are certain characteristics that applicants must possess to be attractive to a police agency, this premise invites two questions:

1. If such characteristics indeed exist, which of them, individually and in combination, are indicative of future ethical behavior?
2. If individuals do exhibit such characteristics, is it because of their moral character, prior life and work experience, education, physical agility, medical history, or age?

Reviewing the selection efforts of police departments in the United States, one might ask these questions:

1. Are there any generalizable evidence-based characteristics that could be looked on as profiling the "ideal" prototype of an ethical police professional?
2. Is the entire selection enterprise highly influenced by a number of themes unrelated to integrity?

Some of these themes stem from political pressure, some from budget constraints, and yet others from a complete ignorance of the posited desired skills and qualities. Profiling is important to consider because frequently individuals who engage in corrupt behavior are associated with various characteristics or lifestyles.

The second theoretical theme is based on the premise that a direct but complex and dynamic relationship between the selection process and integrity can be observed by looking at the planned efforts dedicated by the agency to the selection process. This begins with the allocation of resources and manpower (devoted to the verification of those "minimum standards"), followed by a closer look at the rationale behind the minimum standards required of the potential applicant. These allocations of resources and manpower relate directly both to the role the police play in a given society and/or community and to the culture of policing within the organization on a micro level and within the governmental and societal structures on a macro level. Various types of transformations—economic, societal, political, and cultural—contribute to observable changes in the distribution of resources to the organizations that serve the government and the community.

Training

The component of police training is somewhat more problematic. Police training is defined as a process of socializing a less than homogeneous group of people into police work, during which necessary rules and values of a given organization, system, and culture are communicated to the recruits. Police training, on a basic academy level, is delivered through an in-house police academy, a regional academy, or a state academy. Only departments with their own in-house programs have the ability to invest and maintain real input into the curriculum, number of hours allocated to a given topic, selection of instructors, instructional materials, facilities, etc. Any police academy, regardless of its size, length of existence, or location, serves a certain role in creating and maintaining police integrity. Training, in general, can be defined as a three-step process in which necessary rules and values of police organization are explained, demonstrated, and practiced. In the course of this process, certain situations and behaviors are addressed, and the instructors are responsible for showing the police recruits how to behave and/or respond under a range of challenging and oftentimes complex circumstances and how to do this without sacrificing personal and/or police organizational integrity. This happens almost irrespectively of the quality of the instruction and the instructors. Police recruits are socialized and acculturated to the life of a police officer—getting to work on time, listening to and following instructions, and learning other discipline-related issues as well as being subject to "vulnerability" reinforcing or promoting pressures. These explanations, demonstrations, and practices establish a certain type of an "integrity character."

It is reasonable to assume that there is a general relationship between the number of hours allocated to a given training topic and the issue of police integrity. The longer the period dedicated to the topic, the better is the socialization to the discussed values. Having noted this, the reality is that for any educational and/or training process, it is rare that generalizable evidence-based data exist that form the basis for making staffing and materials choices and decisions as well as making possible the necessary conditions for achieving the selected goals or targets. It is no different for police training.

When looking at the three components of the training process—explanation, demonstration, and practice—we can assess the quality of experience for a given applicant. In the case of integrity/ethics training, the following four steps help to identify the importance of such topics to a given police academy:[1]

1. Examining whether a particular topic is part of the curriculum makes it possible to deduce something about the overall importance of that topic to the given organization at a given point in time or with regard to its future planning. Given the fact that each state mandates the number of hours and the mandatory topics each academy must include in its training, the choice of additional topics is basically an arbitrary decision made by the training director and/or police chief.
2. By reviewing and analyzing how this topic is demonstrated (or not demonstrated) to police recruits, we can begin to see what the academy (or the organization during the on-the-job training) actually trains its "screened-in" recruits to do.
3. We may further engage in investigating the police training process to find out and to better understand what is necessary for a person to know about ethical policing in a democratic society and how to do it before (or after) recruits become police officers.
4. By following the process involved in practicing the policing "gospel," we might arrive at a final conclusion as to whether or not a given topic is actually treated with the ultimate necessary seriousness by the police organization. Looking at the materials used to instruct a given topic, further conclusions might be reached.

No matter how positive the overall conclusion about the triangle of integrity of recruitment, selection, and training, an environment of integrity relies on the two additional prongs of the pentagon: supervision and discipline. Without proper supervision and discipline, it will be impossible to maintain environments of integrity in a world filled with opportunities and temptations to abuse the rights of the position of police officer.

THE PENTAGON OF POLICE LEADERSHIP

As previously discussed, police officers operate in a world full of temptations. Nothing describes it better than a statement made by a representative of the British police force during a meeting sponsored by the Council of Europe in 1999 in Strasbourg, France:

> *Given the nature of police work, it is no shame to find corruption within the service; the shame is not doing anything about it.*

Anti-Corruption Task Force, 1999.
Association of Chief Police Officers, U.K.

Environments of temptation can consist of many prongs and angles. It might be the classic economic gain corruption, the noble cause corruption, or the ambiguity of a law that leaves officers with an impression that discretion is about manipulation and situational integrity, exempt from

---

[1] This chapter does not look at supervision and built-in feedback and their relationship to the three components of training discussed in this segment. Future researchers would be well advised to look at these components as well.

objective evaluation. No matter what the cause, reason, or motivation, the fact remains that police misconduct is rarely tolerated by the public and is certainly not something that should be tolerated by a police leader, no matter how high or low he/she is on the leadership ladder.

It is very hard for this author to make a value judgment about the various situations that require or at least appear to require some manipulation. Nevertheless, the pentagon of police leadership can provide a useful tool for understanding the nature of the problem and how it can be controlled.

It needs to be reiterated that policing is about using ethics and integrity and about differentiating between right and wrong. Police accomplish these ends through their ability and mandate to use coercive force. Force does not have to be just ultimate brute physical force; sometimes just the implied notion of the ability to use this force can be as detrimental. Therefore, no effort should be spared to ensure that the ends are accomplished in the most effective and straightforward manner. Every police officer needs the ability to lead, to take control, and to use force, force that can be lethal if not handled by the right people.

The idea of effective policing will always be a balance between the two prevalent leadership concepts:

1. Doing things right
2. Doing the right thing

A third concept, related to number two, that is of critical importance in police work takes the concept a step further by making it as explicit as it can be and in a provocative way:

3. Sometimes it's right to do the wrong thing

This third concept underscores why developing ethical police leaders cannot be emphasized enough: Individuals who have the mandate to use coercive force against others, and who have the discretion to choose one of these three concepts, need to be very carefully chosen, trained, supervised, and, if needed, disciplined.

This is where one of the leadership theories, presented in more detail later on in this book, becomes more applicable than others where ethical leadership is concerned: the path-goal directive approach (see table in Introduction and Chapter 9). While conducting research in three large police departments in the United States, the author and her colleagues looked at integrity

The Path-Goal Theory of Leadership

1. Subordinates will be motivated if:
   a. They are capable of performing their work.
   b. Their efforts will result in a certain outcome.
   c. The payoffs for doing their work are worthwhile.
2. Leaders help subordinates define their goals and clarify their work.
3. Leaders remove obstacles and provide support.
4. Leaders select a style of leadership that is best suited to their subordinates:
   a. *Directive* leadership gives subordinates instruction about the task (complex).
   b. *Supportive* leadership attends to the well-being and the human needs of subordinates (repetitive).
   c. *Achievement-oriented* leadership establishes a high standard of excellence (excel), consults with subordinates, and integrates their suggestions (control).
   d. *Achievement-oriented* leadership establishes a high standard of excellence (excel).

Five-Step Approach to Integrity Management

Question 1. Do officers in this agency know the rules?

Action Response. If they do, fine. Where they don't, teach them.

Question 2. How strongly do they support those rules?

Action Response. If they support them, fine. Where they don't, teach them why they should.

Question 3. Do they know what disciplinary threat this agency makes for violation of those rules?

Action Response. If they do, fine. Where they don't, teach them.

Question 4. Do they *think the discipline is fair?*

Action Response. If they do, fine. Where they don't, adjust discipline or correct their perceptions.

Question 5. How willing are they to report misconduct?

Action Response. If they are willing, fine. Where they are not, find ways of getting them to do so.

(Klockars et al., 2002, 2006)

management from many different angles. The conclusion reached by the team of researchers produced a five-step approach to integrity management that appears to be replicable in other police environments as well. These five steps are best implemented through the use of the path-goal directive approach to leadership.

The directive approach of the path-goal theory posits that when faced with a complex situation, leaders should give subordinates clear and straightforward instructions about how to complete the task. There is nothing more complex in police work than the issue of police ethics and leadership. While there is room for discussion and dialogue, at the end of the day each and every police officer needs to know what to do and how to do it. Effective supervision and swift and certain discipline will ensure that this knowledge of what and how will not be diluted or otherwise distorted.

If police leaders follow Klockars et al.'s five steps of integrity management, chances are that the complex topic of police ethics will become a tiny bit less complex.

The steps are composed of five questions and answers; readers can customize those questions and answers to the reality of their police organizations and their own status within the organizational structure. Sometimes the nature of the offense or misconduct will allow for a further modification.

Ethical leaders must be developed to handle these complex issues of integrity properly. Policing is about discretion, policing is about use of force, policing is about ethics and integrity—only leaders can handle such heavy ideals and loads. We are not born with the ability to exercise discretion and authority without abusing it and hurting others as a result. The same can be said about the use of force; we do not know how to use it without hurting others. The ability to use force, the ability to use discretion in issues of life and death, the ability to make ethical decisions—those are leadership skills that need to be honed and developed. Hopefully this book will contribute to this process by providing an opportunity to consider the theories, the concepts, and the situational applications.

References

Atwater, L.E., S.D. Dionne, B. Avolio, J.F. Camobreco & A.W. Lao (1999). A longitudinal study of the leadership development process: Individual differences predicting leader effectiveness. *Human Relations*, 52, 1543–1562.

Bass, B & P. Steidlmeier (1999). Ethics, character and authentic transformational leadership behavior. *Leadership Quarterly*, 10, 181–217.

Burns, J.M. (1978). *Leadership*. New York: Harper & Row.

Caldwell, C., S.J. Bischoff & R. Karri (2002). The four umpires: A paradigm for ethical leadership. *Journal of Business Ethics*, 36, 153–163.

Ciulla, J.B. (1995). Leadership ethics: Mapping the territory. *Business Ethics Quarterly*, 5 (1), 5–28.

Crank, J.P. (2004). *Understanding police culture*. Cincinnati, OH: Anderson Publishing.

_____ & M. Caldero (2000). *Police ethics: The corruption of noble cause*. Cincinnati, OH: Anderson Publishing.

Eckhart, D.E. (2003, March). What do ethical leaders follow? The way of truth. *PA Times*, 6.

Giampetro-Meyer, A., S.J.T. Brown, M.N. Browne & N. Kubasek (1998). Do we really want more leaders in business? *Journal of Business Ethics*, 17, 1727–1736.

Greenleaf, R. (1977). *Servant leadership: A journey into the nature of legitimate power and greatness*. New York: Paulist.

Heifetz, R.A. (1994). *Leadership without easy answers*. Cambridge, MA: Belknap Press.

Kanungo, R.N. (2001). Ethical values of transactional and transformational leaders. *Canadian Journal of Administrative Sciences*, 18 (4), 257–265.

Kleinig, J. (1996). *The ethics of policing*. New York: Cambridge University Press.

Klockars, C.B., S. Kutnjak Ivkovich, W.E. Harver & M.R. Haberfeld (1997). The measurement of police integrity. Final report submitted to the National Institute of Justice, NIJ Grant #95-IJ-CX-0058, *The cross cultural study of police corruption*. Also published as the NIJ Research in Brief Report, May 2000.

_____, W. Geller, S. Kutnjak Ivkovich, M.R. Haberfeld & A. Uydess (2002). *Enhancing police integrity: A final report to the National Institute of Justice* (pp. 2–7). Washington, D.C.: National Institute of Justice.

_____, S. Kutnjak Ivkovich & M.R. Haberfeld (2006). *Enhancing police integrity*. New York: Springer Academic Publisher.

Kuhnert, K.W. & P. Lewis (1987). Transactional and transformational leadership: A constructive/developmental analysis. *Academy of Management Review*, 12, 648–657.

Martinez-Carbonell, K. (2003, March). Inching toward ethical leadership. *PA Times*, 7.

Price, T. (2003). The ethics of transformational leadership. *The Leadership Quarterly*, 14 (1), 67–81.

Punch, M. (1999). Police misconduct and "system failure": European perspective. Unpublished paper presented at the International Conference of Police Integrity and Democracies, May 20–23, 1999, Florence, Italy.

Schein, E. (1977). *Organization culture and leadership* (2nd ed.). San Francisco, CA: Jossey-Bass.

Turner, N., J. Barling, O. Epitropaki & V. Milner (2002). Transformational leadership and moral reasoning. *Journal of Applied Psychology*, 87 (2), 304–311.

Weber, J. (1989). Managers' moral meanings: An exploratory look at managers' responses to moral dilemmas. *Academy of Management Proceedings*, Washington, D.C., 333–337.

3

Partnership in a Small Force
Team Theory

OVERVIEW OF THE THEORY

Use of teams in organizations began in the United States in the 1980s. In organizational settings, management creates teams to solve a particular problem, and the teams are temporary in nature. In other settings, teams reflect the primary units within the organization (Dumaine, 1994). In this chapter, the relationship of two officers from the small department of Montclair, New Jersey—Chief Dave Harman and Deputy Chief Frank Viturello—illustrates how the theories of team leadership can be applied in practice in a police environment.

Teams are typically defined as sets of two or more people who cooperate toward a common goal and who have specific functions to perform. Team members are interdependent, coordinated, and synchronized and information exchange is integral to the team's coordination (Salas et al., 1992). Teams are also adaptive in that they can maintain a level of performance even when faced with external pressures (Zaccaro et al., 2001). They also may be able to accomplish tasks that could not be accomplished outside the team environment:

> Team leader effects go beyond those of each individual member interacting with the leader. There may be an assembly bonus effect for the team so that the team does better than the sum of its members. (Bass, 1990, p. 611)

FUNCTIONAL TEAM LEADERSHIP

Functional team leadership considers the use of teams as a form of social problem solving. Effective team leaders diagnose problems, plan appropriate solutions, and implement them (Zaccaro et al., 2001). Most teams originate from a particular environment; therefore, effective leaders are in tune with external events and demands (Ancona & Caldwell, 1988). Leaders interpret and define the outside context for the team and make sense of the cognitive, social, personal, political, technological, financial, and staffing concern factors present in that environment. Thus, the core notion here is that the circumstances surrounding the team prescribe leadership activities. Effective team leaders possess the skills to respond

Effective Team Leaders

- Effective teams often display smooth processes of distributed leadership.
- Groups and teams move through stages of development. Good team leaders understand and manage them.
- Effective team leaders understand the importance of clarifying membership, purpose, and leadership processes early in the team's life cycle.
- Effective team leaders will find ways to revisit the team's vision while it is performing its task to keep energy and motivation high.
- Effective team leaders understand and are able to manage the team's response to changing environmental and internal conditions and events.

(Clawson, 2003)

to and to organize in a given situation. These skills are not a specifically defined set of behaviors (Zaccaro et al., 2001).

Zaccaro et al. (2001) reflects that leadership influences team effectiveness through four types of means: cognitive processes, cohesion, affect, and coordination.

Cognitive Processes

This involves the emergence of shared mental models, which occurs when team members are able to anticipate each other's actions; a feat usually derived from experience, thereby decreasing the amount of planning and communication needed to accomplish a goal. Cannon-Bowers et al. (1993) identified various mental models, including the equipment model (knowledge of the equipment used by the team), task model (tasks, procedures, and strategies), and team model (skills, attitudes, and preferences), and the interplay between the collective and individual needs of the team. According to Zaccaro et al. (2001), the above mental models originating in the leaders are the determinants of the mental model developed by the team, which thereby influences the team's coordination and performance.

As the team matures and the leader has imparted the mental modeling necessary for performance, there can be a shift toward self-management if the leader provides less guidance. With self-managed teams, leaders do not direct team activities; instead, they work on developing the individual and collective capacities of the team to self-manage (Kozlowski, 1998). Likewise, most models of team leadership emphasize participatory leadership which has been shown to be more effective in teams rather than large groups (Bass, 1990). For example, participatory leadership has been shown to increase information sharing and performance of teams (Larson et al., 1998; Tesluk & Mathieu, 1999). Leaders who encourage meta-cognitive practices (reflection on the processes used in problem solving) have also been shown to lead to better team performance than those who do not (Tannenbaum et al., 1998). This suggests that leaders who provide for training, self-critique, and feedback are more likely to have an effective team.

However, an earlier study by Janis (1982) indicates that cohesive groups are susceptible to groupthink, an extreme concurrence in decision making that thwarts realistic evaluation of alternative options and courses of action. Conditions such as directive leadership, undue pressure for an end product, and insulation of the team from external sources of information are likely predicators of groupthink. In another study, the key component in the occurrence of groupthink was not the level of cohesiveness but the existence of a directive leadership style on the part of the leader (Bass, 1990).

Cohesion

Cohesion is what attracts and keeps members in the group as well as how the group resists external disruptions. Zaccaro and McCoy (1988) found that members of teams with high coherence were more likely to be committed to the job, to devote more effort to it, and set higher performance norms. Therefore, creating group cohesion is a way to effectively lead teams; however, groups must work together for a long time for cohesion to develop (Bass, 1990). In a study of 203 military cadets, cohesiveness moderated the effect of leaders' consideration and initiation of structure on subordinate teams (Bass, 1990; Dobbins & Zaccaro, 1986).

According to Zaccaro et al. (2001), a common theme in the literature on team leadership is the importance of collective efficacy as originally conceptualized by Bandura. This refers to members' beliefs that they are capable of accomplishing their objectives. When teams have high collective efficacy, they tend to work harder in the face of challenges. Zaccaro et al. (2001) argues that the leaders' role as builders of collective efficacy is similar to the empowerment process characterized by transformational leadership.

Affect

Leaders must also control the emotional affect of the team because collective negative moods lead to more internal conflict and less motivation to complete tasks:

> Teams performing under stressful conditions can be highly susceptible to emotional distress across members. As team environments become more aversive . . . , team members obviously need to maintain a collective calm. If the team succumbs to stress, member interactions become more and more narrowly focused among a subset of the team, information becomes increasingly less shared among team members, decision alternatives are not fully explored, and decision making accuracy declines. (Zaccaro et al., 2001, p. 471)

Studies have shown that teams suffer from less emotional reactions to stress when leaders provide clearly outlined objectives, job descriptions, and performance strategies (Isenberg, 1981; Sugiman & Misumi, 1988). Emotional affect can also be controlled by creating a climate in which conflict is handled constructively. In essence, the leader develops the interaction patterns necessary for team performance (Zaccaro et al., 2001).

Coordination

Effective team leaders help their team members coordinate actions. This is accomplished by setting up orientation functions, distributing resource information, working on timing functions (pace and speed of task), working on detecting errors (system monitoring), and making sure operations conform to established standards (procedure maintenance; Zaccaro et al., 2001). As Kozlowski et al. (1996) points out, when the complexity of the demands on the team grows, established interaction can become insufficient. Leaders can adapt to this by reorienting the team to new patterns and by encouraging flexibility among individuals on the team, as noted by Zaccaro et al. (2001).

THE LEADER-MANAGEMENT MODEL

Here, leadership is broken down into three categories: conceiving, organizing, and accomplishing. Ideas are generated and plans are made (conceiving phase). A strategy is then designed to implement the plan (organizing phase), which is carried out through a series of events (accomplishing stage) whose aim is the achievement of the objectives of the project. Weinkauf and Hoegel (2002) findings also support situational leadership, in that types of leadership activities were performed when needed, depending on the stage the team was in, rather than being consistently performed over the life of the team.

SELF-MANAGED TEAMS

Decision-making authority on these teams, which usually consist of between 8 and 15 members, rests with the individual team members. Here, team members are involved in planning, measuring, and executing entire operations (Jessup, 1990). Leadership is often rotated amongst team members so that each member feels that he/she plays a part in the team's overall performance (Erez et al., 2002). Rotated leadership has been compared to that of club officers in which everybody is a leader (Jessup, 1990). Kerr (1983) hypothesizes that rotated leadership leads to a greater likelihood that team members will pull their weight in accomplishing tasks.

Another form of leadership present in self-managed teams is emergent leadership, in which individuals step forward to take the leadership role. Some theorists believe that a natural selection process occurs in small groups from which the most qualified leader organically emerges (Erez et al., 2002). In a quasi-experiment testing the effects of peer evaluations and leadership in self-managed teams, Erez et al. (2002) reports that differences in team design impact effectiveness. Specifically, the use of peer ratings on teams promotes workload sharing and cooperation, which all lead to higher team performance. Similarly, rotated leadership promotes cooperation, enhancing team performance.

TEAM POLICING

In the 1970s, team policing was practiced in some jurisdictions in an attempt to transition toward community-oriented policing. In its idealized form, it is a flexible participatory structure in the form of a self-managed team with rotated leadership in which administrative procedural guidelines do not apply to team members and goals are set at team meetings, with direct input from citizens. However, team policing was not successful for several reasons including the fact that the team tended to be implemented in the midst of the existent hierarchical structure of policing, excluded middle managers sabotaged team policing efforts in their departments, and the perception that team policing could lead to corruption amongst patrol officers (Toche & Grant, 1991).

Brewer et al. (1994) studied a sample of Australian police teams and showed that monitoring team performance and soliciting self-reports lead to higher performance. Using a sample of 20 police patrol sergeants, each of whom supervised a stable team of officers, the researchers analyzed the sergeants' performance evaluations (written by their superiors) as well as directly observed the sergeants with their teams, evaluating whether each sergeant communicated expectations and consequences. The researchers found that sergeants spend 53% more time on average supervising subordinates than supervisors in insurance, newspaper, and bank work environments. Even though inadequate feedback characterizes most police organizations, sergeants' use of performance monitoring and consequences was strongly correlated with effective team performance. This suggests that the team model may be compatible with the quasi-military chain of command if implemented properly.

Nahavandi (2003) differentiates between two crucial concepts—teams and groups (see Table 3-1). This differentiation clearly points to the fact that in police environments it will be very hard to find a real team (as defined by Nahavandi); instead, group environments are primarily encountered. The five ingredients necessary to create a true team will occur very rarely, precluded

TABLE 3-1	Groups and Teams	
Groups	**Teams**	
Work on common goal	Are fully committed and develop a mission	
Accountable to manager	Mutually accountable	
Conflict—no clear culture	Trust—collaborative culture	
Leadership assigned to one person	Shared leadership among members	
May accomplish their goal	Achieve synergy ($2+2=5$)	

(Nahavandi, 2003)

by the organizational and cultural or subcultural structure. It is very rare to find a team whose members are fully committed to developing a mission of police organization. Accountability is usually individual, not mutual. Trust or a collaborative culture is frequently contradictory to police subculture, which more often emphasizes a suspicious and dubious view of the world. Leadership sharing is individual rather than delegated (in the full sense of the concept). Similarly, leadership is routinely associated with higher rank, and an officer with higher rank is the one who sets the final perimeters to any kind of empowerment. Finally, achievement of true synergy can only be accomplished when one steps out of the box, not just in symbolic terms but with the risk-taking approach, and risk taking in police environments is only encouraged within certain clearly defined and liability-oriented parameters.

On the other hand, working on common goals, being accountable to one's supervisor, being exposed to an esoteric subculture that is unique and that differs between units and teams, and experiencing the overall environment of conflict in which one may or may not accomplish a goal (formal or informal) appear to characterize many, if not all, police organizations, regardless of their size and location. These elements lend themselves to creating environments destined for group rather than team orientations.

CAREER HIGHLIGHTS

Team Leadership Approach

Chief David Harman and Deputy Chief Frank Viturello
Montclair Police Department, New Jersey

One cannot completely exclude the possibility of finding a true team approach in police organizations, and an example of such collaboration can be found in the very real, and rare, cooperation between Chief David Harman and Deputy Chief Frank Viturello of the Montclair Police Department in New Jersey.

Most of the police chiefs depicted in this book come from large and/or notorious police departments. This choice was not coincidental. It is much easier to collect information about police chiefs and departments that are large and well known, particularly if there has been a scandal or misconduct. The nature of events, a geographic location, and its prominence or lack of prominence in the daily media will dictate the level of coverage that will, in turn, be available to researchers. Small police departments also frequently undergo interesting challenges and exhibit leadership characteristics worthy of wider dissemination. It is a rare opportunity to get insight into a smaller police department where things and events are known primarily to the local community and written about only in the local newspapers.

The partnership between Chief Harman and Deputy Chief Viturello exemplifies, as closely as possible, the five ingredients of the team approach to leadership (as defined by Nahavandi, 2003). Although their police careers and personal backgrounds did not necessarily make them an obvious match, their dedication to the police profession, their basic values, and their ethical consciousness created a synergy that can and should be looked upon as a model approach to team leadership in police environments. To fully understand the uniqueness of this partnership, it is necessary to follow their careers on the force, on an individual level, and identify the traits and qualities that brought them together and that allowed them to experience the level of comfort necessary for the implementation of the team leadership approach. Of course, this summary only highlights certain aspects of their relationship and careers—the choice of these was intentional to illustrate the team leadership theory.

Montclair, New Jersey, is a township of approximately 39,000 people that prides itself on its diversity; approximately 59% of the population is white, 32% is African American, and the majority of the remainder of the minority population is Latino or Hispanic (U.S. Census, 2000).

The Montclair Police Department had 111 sworn members in 2004, of whom 28% were minorities or women. Harman served in the Montclair Police Department for 25 years, retiring in May 2004 after serving 3 years as chief (Zadrozny, 2004; Brubaker, 2004). An African American whose family moved to Montclair when he was in high school, Harman served as a military police officer before joining the Montclair Police Department in 1978 (Harrell, 2001a).

Harman was an early rising star, ascending quickly through the ranks. He made detective 28 months after joining the force in 1979. He spent more than seven years as a detective and then was promoted after scoring second on the sergeant's promotional exam. He was first put in charge of the patrol division and later put in charge of the detectives. After 13 years on the job, he became a lieutenant and within seven months was promoted to deputy chief. In 2001, he was promoted to chief, the first African American appointed to that position in the history of the force.

Viturello retired from the force as deputy chief in 2003. An Italian American, he joined the Montclair Police Department in 1970, after graduating from the police academy first in his class. A union leader, Viturello views one of his accomplishments as successfully advocating for the introduction of a merit-based promotional system within the force, introduced in the 1980s. It was only after this system was put in place that Viturello's career advanced; the prior system had relied on political favoritism that excluded Viturello, partially due to his ethnicity and his outsider status (he did not live in Montclair). After the introduction of the system, he gradually made his way up through the ranks. Although formally he became part of a senior management team with Harman when Harman appointed him deputy chief in 2001, they had a close working relationship from long before that time, when Viturello served as Harman's sergeant when Harman made detective.

Harman and Viturello both describe their working relationship as "being on the same page." They held similar goals for the department: to increase its diversity and equitability. Before they assumed positions of leadership, the department had been run according to a "good old boys' network" (D. Harman, personal interview, June 20, 2004). Both Harman, as an African American, and Viturello, as an Italian American, had experienced discrimination early in their careers. Viturello, as a union member, advocated and was instrumental in stating a merit- and examination-based system of promotion and career advancement. Harman increased the level of minority representation on the force through a commitment to recruitment he describes as "a continuous job. You have to be highly motivated. It will take a lot of time, and the chief of police has to be 100% behind that."

The recruitment policy change was one of the first major changes brought in by the department after Harman became chief. When Harman took over the force, less than 20% were minorities or women (Harrell, 2001b; Thorbourne, 2003). With town officials, he changed recruitment policies so that Montclair residents were given priority on the recruitment test. Given the ethnic makeup of the township and the fact that 92% of those who took the first test offered were minorities, the numbers of minorities increased significantly (Thorbourne, 2001).

Montclair also was the first department in the state to implement a policy designed to monitor racial profiling; video cameras were installed in patrol cars, and information on the race, ethnicity, and gender of those stopped was logged and recorded statistically (Harrell, 2001c). Eighteen months after the policy was implemented, results showed that the racial makeup of those stopped reflected the township's ethnic makeup (Thorbourne, 2002).

During the period when Harman and Viturello led the department, there were several scandals. A police officer was accused of beating an epileptic man and contributing to a seizure episode (Harrell, 2001d). Another officer resigned after being accused of sexually assaulting a woman whom he had followed back to her apartment after she had no documents during a traffic stop (Harrell, 2001d).

Harman and Viturello describe the key aspects of their management and leadership in similar terms: a commitment to demonstrating integrity through example, treating all employees fairly, and motivating people to work by making them want to do the work rather than forcing them to do it. This shared vision was one that was present near the beginning of their relationship. Viturello describes their relationship as developing over a number of years but that the shared vision they had was something innate to both of them; it was the common

bond they shared through these values that strengthened their partnership.

As an example of their shared vision, they had similar approaches to managing subordinates when subordinates' requests (e.g., for time off) could not be fulfilled; both preferred the style of explaining the reasons and justifications for the denial rather than simply playing the authority card. Viturello gave one such example:

For example, the chief says, "Look, we've got to cut back on vacation, so get guys to keep guys on deck." A weak guy would go the squads and say, "It's not me. The chief said we have to do it. It's not me." Me, I would say, "Guys, we have a problem; we have a crime wave. You've got to work with me. If you ask for time off, I'm not going to be able to give it to you now. If you work with me, I'll be there for you later." That's being honest. Most of the time they appreciate it. (Personal interview, June 16, 2004)

Harman gave an example illustrating the same approach:

When I took over, there was a big problem with overtime. We had to get people to seriously look at staffing. From 1993 to 1999, the department had received over 3 million dollars in grants, and they hired more police officers. But the grants were drying up. But the [community] had expectations, expected a certain level of service. And supervisors expected to be able to provide that level of service. But there was no money for it. So people had to know why, and I'd explain it to commanders. I'd look over the schedules and say, "Look, you hired two people to work this shift. Why?" So I started on top and worked my way down through the ranks, to get everyone on the same page. (Personal interview, June 20, 2004)

Furthermore, both Harman and Viturello emphasize trust and collaboration in their approach not only to their subordinates but also to each other. There can be no doubt that the partnership of Harman and Viturello was a tight unit. Viturello describes their relationship in emotional terms: "His love and respect for me, and my admiration for him, that's what made our relationship tight" (Personal interview, June 16, 2004). Harman describes Viturello as someone who "always had my back" (Personal interview, June 20, 2004).

Although not exactly key to the team itself, motivation was a key element of the team's leadership style for the force. Harman and Viturello both emphasize that believing in their subordinates was key to motivating them. Harman used the key term "empower":

As for my style, what I do is empower my personnel. I involve them in the decision-making process; I know I can't do it alone. We need to be on the same page, and I let them know that. It makes for a better department. You can't make someone follow you, but they will do it if they want to follow you. (Personal interview, June 20, 2004)

Viturello used similar language, illustrating with an example of an officer to whom he gave a second chance:

For example, one guy was on the border for abusing sick time, and I had even scolded him for a poor traffic report. He knew he was in trouble, and I found an opportunity for him. I said I need someone to review reports. Why? I knew he had ability; I knew he was intelligent. I saw some of myself in him, and I saw that the difference was because I had the urge to fight. Certain things I saw, I would fight them and go to the chief. I'm not afraid to reward those who needed to be rewarded. Not afraid to stand up for those who needed to be stood up for. . . . I felt I had to motivate. I tried not to raise my voice, but to do it through discipline. People will work when they want to, not when they have to.

Somewhere I got this principle. I wanted to have people want to work for me. (Personal interview, June 16, 2004)

The two were also unique in their willingness to take risks in order to solve problems. For example, there was an explosive incident in the community when gang members wanted to erect a memorial for one of their members after a shooting (this followed in the wake of a large community memorial for a youth who had died after pushing his girlfriend out of the way of an oncoming train). Viturello described how the two found a solution to placate the competing demands of the members of the community (who did not want to be seen rewarding gangs) and the gang members (who might take action that would lead to further gang violence if the tensions were not resolved):

The community didn't want this; there was a lot of pressure to say no. We could have said that graffiti is defacing public property. We went down and spoke to the gang leaders and said that it was defacing property, etc. They said, "If you leave us alone, after two or three days, we'll clean it up." I said that it was better to have two to three days of graffiti than months of cop overtime. (Personal interview, June 16, 2004)

Harman also described the reaction in the community and his response:

And you have to see the whole picture. We had intelligence that said that if we stepped in and tore it down, it would be trouble. We even talked to some of our informants in town, and they said the same. I hired more cops on overtime [for the memorial service]; they kept an eye on things. In the end, there was no violence. We knew, on the other hand, that if we stepped in, it would be violent. . . . Well, you have to understand; the community, they understood. Again, the problem was something in the media. People, they understand; they have to live there. People in gangs live there too, and they have to live next door [to them]. Do you really want to provoke problems? On the other hand, if you respect them, look, they came in and cleaned it up afterwards. That's what I mean. Sometimes you have to take a stand. You have to look at the whole picture and understand what would give you the most serious repercussions. (Personal interview, June 20, 2004)

As a result of the willingness of Harman and Viturello to consider a creative approach, the memorial was peaceful, and the gang members cleaned up the area of the memorial, resulting in no further public criticism or outcry.

Exercise

The relationship between Dave Harman and Frank Viturello (the police chief and his deputy from Montclair, New Jersey) detailed in this chapter has been examined through the prism of team leadership theory. Using what you have learned from other theories in this book or from the summaries (as indicated in the Introduction) you have read thus far, it is possible to consider their approach through the lens of different theories. What other theories can you find that are applicable to these chiefs? Are there lessons to be learned from those applications, or do those theories indicate alternative paths or approaches that might have yielded a more successful outcome?

References

Ancona, D.G. & D.F. Caldwell (1988). Beyond task maintenance: Defining external functions in groups. *Group and Organization Studies*, 13, 468–494.

Bass, B.M. (1990). *Bass and Stogdill's handbook of leadership: A survey of theory and research.* New York: Free Press.

Brewer, N., C. Wilson & K. Beck (1994). Supervisory behaviour and team performance amongst police patrol sergeants. *Journal of Occupational and Organizational Psychology,* 67, 69–78. *Personnel Psychology,* 55, 929–948.

Brubaker, P. (2004, May 19). May the force be with them: Sabagh named new chief of police, Terry is new deputy. *The Montclair Times.* Retrieved July 22, 2004, from http://www.montclairtimes.com/page.php?page=7616.

Cannon-Bowers, J.A., E. Salas & S. Converse (1993). Shared mental models in expert team decision making. In N.J. Castellan, Jr. (Ed.), *Current issues in individual and group decision making* (pp. 221–246). Hillsdale, NJ: Lawrence Erlbaum.

Clawson, J.G. (2003). *Level three leadership: Getting below the surface.* Upper Saddle River, NJ: Pearson Education, Inc.

Dobbins, G.H. & S.J. Zaccaro (1986). The effect of group cohesion and leader behavior on subordinate satisfaction. *Group and Organization Studies,* 11 (3), 203–229.

Dumaine, B. (1994). The trouble with teams. *Fortune,* 130 (5), 86–92.

Erez, A., J.A. Lepine & H. Elms (2002). Effects of rotated leadership and peer evaluation on the functioning and effectiveness of self-managed teams: A quasi-experiment. *Personnel Psychology,* 55, 929–948.

Harrell, J. (2001a, February 2). Russo to have respite before taking new post. *The Montclair Times.* Retrieved July 22, 2004, from http://montclairtimes.com/page.php?page=1872.

_____ (2001b, February 22). Montclair aims to diversify police force. *The Montclair Times.* Retrieved July 22, 2004, from http://www.montclairtimes.com/page.php?page=1882.

_____ (2001c, March 1). New policy announced to track police racial profiling with statistics. *The Montclair Times.* Retrieved July 22, 2004, from http://www.montclairtimes.com/page.php?page=1884.

_____ (2001d, May 24). Cop resigns and avoids charges. *The Montclair Times.* Retrieved July 22, 2004, from http://www.montclairtimes.com/page.php?page=1929.

Isenberg, D.J. (1981). Some effects of time-pressures on vertical structure and decision-making accuracy in small groups. *Organization Behavior and Human Performance,* 27, 119–134.

Janis, I.L. (1982). *Groupthink* (2nd ed.). Boston, MA: Houghton Mifflin.

Jessup, H.R. (1990). New roles in team leadership. *Training and Development Journal*, 44, 79–83.

Kerr, N.L. (1983). Motivation losses in small groups: A paradigm for social dilemma analysis. *Journal of Personality and Social Psychology*, 45, 819–828.

Kozlowski, S.W.J. (1998). Training and developing adaptive teams: Theory, principles and research. In J.A. Canon-Bowers & E. Salas (Eds.), *Making decisions under stress: Implications for individual and team training* (pp. 115–153). Washington, D.C.: American Psychological Association.

Larson, J.R., Jr., P.G. Foster-Fichman & T.M. Frantz (1998). Leadership style and the discussion of shared and unshared information in decision-making groups. *Personality and Social Psychology*, 24, 482–495.

Montclair, New Jersey (n.d.). Retrieved July 22, 2004, from http://www.city-data.com/city/Montclair-New-Jersey.html.

Nahavandi, A. (2003). *The art and science of leadership* (3rd ed.). Upper Saddle River, NJ: Prentice Hall.

Salas, E., T.L. Dickinson, S.A. Converse & S.I. Tannenbaum (1992). Toward an understanding of team performance and training. In R.W. Swezey & E. Salas (Eds.), *Teams: Their training and performance* (pp. 3–29). Norwood, NJ: Ablex Publishing.

Sugiman, T. & J. Misumi (1988). Development of a new evacuation method for emergencies: Control of collective behavior by emergent small groups. *Journal of Applied Psychology*, 73, 3–10.

Tannenbaum, S.I., K. Smith-Jentsch & S.J. Behson (1998). Training team leaders to facilitate team learning and performance. In J.A. Cannon-Bowers & E. Salas (Eds.), *Making decisions under stress: Implications for training and simulations* (pp. 247–270). Washington, D.C.: American Psychological Association.

Tesluk, P.E. & J.E. Mathieu (1999). Overcoming roadblocks to effectiveness: Incorporating management of performance barriers into models of work group effectiveness. *Journal of Applied Psychology*, 84, 200–217.

Thorbourne, K. (2001, August 16). Homegrown recruits enhance police diversity. *The Montclair Times*. Retrieved July 22, 2004, from http://www.montclairtimes.com/page.php?page=2280.

_____(2002, November 29). Civil Rights Commission gives housing crisis a look-see. *The Montclair Times*. Retrieved July 22, 2004, from http://www.montclairtimes.com/page.php?page=4169.

_____(2003, February 26). Rights board looks at police hiring and student gap. *The Montclair Times*. Retrieved July 22, 2004, from http://www.montclairtimes.com/page.php?page=4742.

Toche, H. & J.D. Grant (1991). *Police as problem solvers*. New York: Plenum Press.

Weinkauf, K. & M. Hoegl (2002). Team leadership activities in different project phrases. *Team Performance Management: An International Journal*, 7/8, 171–182.

Zaccaro, J.J. & M.C. McCoy (1988). The effects of task and interpersonal cohesiveness on performance of a disjunctive group task. *Journal of Applied Social Psychology*, 18, 837–851.

Zaccaro, S.J., A.L. Rittman & M.A. Marks (2001). Team leadership. *Leadership Quarterly*, 12, 451–483.

Zadrozny, A. (2004, March 10). Chief retiring after 25 years on the force: David Harman moving to Georgia. *The Montclair Times*. Retrieved July 22, 2004, from http://www.montclairtimes.com/page.php?page=7166.

4

In-Groups and Community-Oriented Policing
Leader–Member Exchange Theory

OVERVIEW OF THE THEORY

The leader–member exchange theory can be understood via the examples of Commissioner Lee P. Brown (Chief-Houston Police Department; Commissioner NYPD) and Chief Darrel Stephens (Chief-Charlotte-Mecklenburg Police Department). Their careers and their relationships with in-groups and out-groups illustrate some of the key principles of the theory.

According to this theory, leadership does not necessarily involve leaders treating each of their subordinates in the same fashion.[1] Leaders instead form different relationships with different subordinates known as vertical linkages (Liden & Graen, 1980). Here, the unit of analysis is the dyad—the employee/manager relationship (Schriesheim et al., 2001).[2] A study by Liden and Graen (1980) found that of 41 dyads, 90% exhibited superiors who had differential relationships with subordinates.

Initially, this theory involved the formation of two groups of subordinates under the leader: the in-group and the out-group (see Table 4-1). In-group employees enjoyed a high-quality relationship with the manager, characterized by more influence over decisions, extra benefits, and greater status, in exchange for higher expectations, task commitment, and loyalty. Contrastingly, out-group employees simply performed their formal job requirements and had a low-quality relationship with the leader (Graen & Cashman, 1975). According to Graen and Cashman (1975), a middle group falls between the in- and out-groups, forming an exchange trichotomy of in-groups, out-groups, and middle groups. In addition, supervisors have high quality relationships with only a few subordinates in the in-group because it is more efficient for them. However, Graen and Uhl-Bien (1991) state that managers should form high-quality relationships with all of their subordinates, thereby making everyone a part of the in-group. According to Graen (1983), by improving the quality of dyadic relationships with employees, managers can increase employee productivity.

---

[1] According to LeBlanc et al. (1993), the assumption that leaders have a uniform way of treating all their employees may be a key reason as to why studies of leadership style have had inconsistent results.

[2] Schriesheim et al. (2001) have shown that there is a statistical problem with all of the work testing leader–member exchange theory: Researchers have used an individual unit of analysis when a dyadic one is appropriate. They suggest, therefore, that the stream of research is fundamentally flawed and that little is actually known about the theory's applicability.

TABLE 4-1	Dyad as Analysis Unit		
In-Group		**Middle Group**	**Out-Group**
High-Quality Relationships			*Low-Quality Relationships*
• Have more influence over decision making • Have extra benefits and status			• Merely fulfill job requirements • Receive the standard benefits

(Graen & Cashman, 1975)

More recently, theorists have posited that each dyadic relationship is unique and that there does not necessarily have to be an out-group or middle group. This individualized approach emphasizes individuals and explores the differential levels of empowerment per subordinate under a particular leader (Yammarino et al., 2001),[3] suggesting a circumstantial continuum of relationships stretching from in-relationships to out-relationships.

MEASUREMENT OF THE QUALITY OF THE DYADIC RELATIONSHIP

The quality of leader–member exchange is measured via the Leader Member Exchange (LMX) instrument (Northouse, 2004). Early studies of this theory used a median-split procedure to define who was in the in-groups (i.e., LMX above median) and out-groups (i.e., LMX below median). Some scholars criticized this approach arguing that the quality of dyadic relationships should be considered a continuous variable (LeBlanc et al., 1993; Vecchio & Gobdel, 1984; Yammarino et al., 2001). A higher-quality LMX score reportedly correlated with more job satisfaction, less turnover, a more positive review of the environment of the organization in general, and salary raises and promotions (Erdogan et al., 2002; Kozlowski & Doherty, 1989; LeBlanc et al., 1993; Vecchio & Gobdel, 1984).

The LMX instrument provided researchers with a composite score summarizing the quality of the dyadic relationship. LeBlanc et al. (1993), however, believed that two aspects of the relationship should be disaggregated: 1) personal attention to the employee and 2) assistance for the leader by the employee. However, Norris and Vecchio (1992) disagreed, arguing that these two categories are synonymous with leadership style and situational approaches. Regardless, a 1993 study by LeBlanc et al. found two things: 1) that the disaggregated factors were both positively related to the quality of the leader–member exchange and 2) that more empirical evidence supporting this theory can be garnered if the LMX score is not considered unidimensional.

THE DYADIC RELATIONSHIP AND EFFECTIVENESS

Liden and Graen (1980) link high-quality dyadic relationships to increased effectiveness, operationalized as high-performance ratings of subordinates by leaders.

Results from other studies are mixed regarding the correlation between the quality of the relationship and job performance (Dunegan et al., 1992).

According to Dunegan et al. (1992), the relationship between the two variables can be predicted only after introducing the moderating variables of task analyzability and variety. In

---

[3] It is noted, however, that Yammarino et al. (2001) do not merely discuss the leader–member exchange theory but place it within the framework of a multidimensional approach that includes style theory and team leadership.

TABLE 4-2	Quality of Exchange	
	Yes	**No**
Routine/Unchallenging Task	Quality of performance is based on LMX.	No relationship exists between LMX and performance.
High Degree of Uncertainty	Quality of performance is based on LMX.	No relationship exists between LMX and performance.

(Dunegan et al., 1992)

their study, they measured subordinates' job performance as the performance rating given by supervisors.[4] Task analyzability was operationalized as the degree to which the subordinate's task involved a particular process or whether the task can only be accomplished through creative problem solving. Task variety refers to the degree to which the subordinate's task is routine or repetitive. What they found was that when employees worked on tasks that were routine and unchallenging, or when the tasks had a high level of uncertainty, the quality of their performance was based on the quality of the leader–member exchange. When the tasks had a more moderate level of uncertainty and challenge, there was no relationship between the leader–member exchange and performance (see Table 4-2). This suggests that the leader–member exchange theory could benefit from the addition of moderating agents, which would develop more of a contingency approach to the dyadic relationships (Dunegan et al., 1992).

Communication researchers have used this theory to explore how workplace communication patterns reflect the quality of the dyadic relationships. Yrle et al. (2003) explored communication style relating to three factors (degree of coordination, participation, and freedom of expression) in the relationship and found that relationships that scored high on the leader–member exchange scale were correlated with higher scores on the degree of coordination of work activities and participation of the subordinate in decision making. In contrast, low-quality relationships were correlated with only a high degree of coordination of work activities via effective communication. Likewise, in low-quality relationships, there may be coordination of activities through effective communication, but this does not lead to true participation in decision making, suggesting that participation is a key component for a healthy relationship (Yrle et al., 2003).

Maurer et al. (2002) explored whether subordinate-perceived level of organizational support and the leader–member exchange dyadic relationship acted as a backdrop to the success of the subordinates' overall career development. Thus, leader–member exchange was treated as one antecedent to employee development. Maurer et al. (2002) proposed that when the dyadic relationship or the perceived level of organization support is weak, only personal benefits (e.g., salary) will motivate subordinates, limiting their career development and work activities. Ideally subordinates would be motivated by benefits both for themselves and for the organization. Maurer et al. (2002) also suggest the prescriptive notion that managers must tap into each employee's individual talents and work styles in order to draw out loyalty and enthusiasm for the organization as a whole and not merely be satisfied with employees who work entirely for their own self-interest.

---

[4] One of my doctoral students suggested that this can represent a tautology. If the dyadic relationship is positive, this could be because the subordinate performs well, necessitating a good evaluation, and the positive relationship may not necessarily be the reason the subordinate performs well. In other words, the independent and dependent variables may be measuring the same thing.

ANTECEDENTS TO HIGH-QUALITY DYADIC RELATIONSHIPS

Leader–member exchange theorists emphasize that a key component in the quality of the exchange between employee and subordinate is the degree to which the employee has influence over organizational decisions.

In essence, Scandura et al. (1986) consider decision-making influence as an antecedent to effective leader–member exchanges finding that subordinates who feel that they have a high degree of influence over decisions (approaching delegation on the continuum) also have higher-quality leader–member exchanges, a phenomenon that occurs regardless of the superior's ratings of the subordinate's performance. However, where leader–member exchange is weak, superior ratings are critical to subordinates' performance. Conversely, managers are most satisfied with employees with whom they share a high-quality leader–member exchange and for whom they give strong performance ratings (Scandura et al., 1986). Likewise for the dyadic relationship to work, subordinates need either informal positive feedback in the form of shared decision making or the formal affirmation of a positive performance evaluation. In turn, superiors need to foster relationships with employees, yet they also need to surround themselves with high-performance employees.

Another antecedent factoring into high-quality relationships is the degree to which the leader and subordinate share demographic characteristics and values. Duchon et al. (1986) found that gender matching partially explained a high-quality exchange. Steiner (1988) also adds that high-quality relationships are correlated with perceived similarity of values, as evaluated by subordinates suggesting that leaders could emphasize commonalities with subordinates in order to build a high-quality relationship.

LEADER–MEMBER EXCHANGE THEORY AND BIAS

An ethical question has surfaced as to whether it is acceptable for a leader to treat employees differently and whether this theory actually describes/promotes discrimination. Research suggests that designation to in-groups and out-groups occurs early in the life of a particular dyad and may be resistant to change (Dinesh & Liden, 1986; Graen & Cashman, 1975). Likewise, some theorists

High-Quality Dyadic Relationships

1. *Autocratic.* The leader makes decisions without the input of the subordinate.
2. *Minimal involvement.* The leader asks for the subordinate's opinion but makes the decision on his/her own.
3. *Consultation.* The leader meets with the subordinate and seeks information, ideas, and suggestions, and incorporates the subordinate's input into the decision making as much as possible.
4. *Collaboration.* The leader and subordinate analyze the problem together, and there is near equal input into the process by both.
5. *Delegation.* The leader permits the subordinate to make the final decision after sharing his/her perspective on the problem.

High quality: Shared decision making or formal affirmation of a positive performance evaluation.

(Scandura et al., 1986)

have posited that the quality of the relationship is based on demographic similarities rather than on any objective measure of subordinates' capacity in the workplace (Yrle et al., 2002). Yet at least three studies have reported that demographic similarity (race, gender, age) does not influence the effectiveness of dyadic relationships (Steiner, 1988; Yrle et al., 2002, 2003). However, Duchon et al. (1986) findings say that it does.

LMX THEORY AND THE CASES OF COMMISSIONER LEE P. BROWN AND CHIEF DARREL STEPHENS

This theory highlights the essence of human relationships; specifically, the fact that we do not treat everybody the same way. We have our favorite people, those to whom we are indifferent, and those whom we dislike. We are also somebody's favorite person, some people are indifferent toward us, while others don't really like us, and sometimes someone really, really dislikes us. However, it is not possible to elevate oneself above these feelings and emotions and treat everybody equally. It is also possible to interact, trust, and support people equally in a work environment. Consequently, all work environments involve in-group and out-group relations. In police environments, where emotional and physical trusts are of paramount importance, this classification into in- and out-groups is more pronounced.

Police work is about reliance on others. It cannot be a one-man or one-woman operation. This does not mean that it will be easy or even possible to identify a true team environment in police organizations (see Chapter 3); however, the prevalence of the in- and out-group phenomenon is something that makes the subculture of policing so different from other organizational subcultures. From both the vantage point of policing as the vanguard of a democratic society and the perspectives of individual police officers on the beat, members of the public, police chiefs, and local politicians, the stakes in police work are very high. When the stakes are high, so are emotions. One of the better-known aspects of police subculture is the prevalence of suspicion and very limited trust. To trust one's subordinates, one must know them quite well and have a common history; to gain trust or maintain it, one frequently must be prepared to compromise principles.

Graen and Uhl-Bien (1991) posit that effective leaders should develop in-group exchanges with all subordinates if possible. In police environments, several prevailing factors make this impossible. To begin, policing is highly politicized and police departments are held accountable by civilian bodies for their misdeeds. Also police unions are traditionally at odds with the police bureaucracy; specifically, police chiefs and commissioners. This tradition exists in all police organizations regardless of shape and size. The cases of the two police chiefs discussed below focus on their inability to include unions in the in-group category, leading to criticism of their leadership as partially ineffective, however untrue or unfair this characterization of their abilities is.

It is simply impossible to make everybody a part of one huge in-group. Clashes with the out-group members are a reflection of reality, not a sign of poor leadership. The way one creates in-groups and maintains them and the way one deals with the out-groups are a true measure of successful leadership when one uses the LMX approach to police leadership. The two chiefs profiled below, Commissioner Lee P. Brown and Chief Darrel Stephens, illustrate these difficulties, particularly with regard to implementation of community-oriented policing. They both have long illustrious careers; the examples and situations described here are merely highlights to illustrate the in-group/out-group concept and how it applies to policing.

CAREER HIGHLIGHTS

Commissioner Lee P. Brown
Houston Police Department, Texas, and New York Police Department, New York

Lee P. Brown was born in Wewoka, Oklahoma, in 1937 (Belkin, 1989a). He received a bachelor's degree in criminology from Fresno State University in 1961, a master's degree in sociology from San Jose State University in 1964, and a doctorate in criminology from the University of California Berkeley in 1970 ("Biography," n.d.). In 1972, he became an associate director of the Institute of Urban Affairs and Research at Howard University, Washington, D.C. (Belkin, 1989a).

Brown entered policing in 1960 as a patrol officer in San Jose, California. In 1968, he relocated to Portland, Oregon, where he founded the Department of the Administration of Justice at Portland State University. In 1975, he served as sheriff of Multnomah County, Oregon. The following year he became the director of justice services, heading the county's criminal justice agencies ("Biography," n.d.).

In 1978, Brown became the public safety commissioner in Atlanta, Georgia, where he oversaw the Atlanta Police and Fire Departments and the Departments of Correctional Services, Emergency Management Services, and Taxis and Limousines. He was noted for starting neighborhood policing in the city by setting up a number of police substations (Belkin, 1989a). In addition, as public safety commissioner, Brown led the investigation into Wayne B. Williams, a serial killer who killed 28 young African Americans; establishing a special task force to handle the case (Harris, 1982). It was this investigation that earned Brown a national reputation as an effective law enforcement leader (Balz, 1982; Balz & Harris, 1982). He was nicknamed "No Rap" because of the limited amount of information he gave to the public during the investigation (Kurtz, 1989).

In 1982, Brown became the first black chief of the Houston Police Department, where he served until 1989. At that time, the force numbered 3,200 officers (Balz, 1982). The force complained that it was spread thin due to population growth and the large geographic area over which Houston is stretched (Balz & Harris, 1982). Brown inherited a department plagued with low morale and a reputation for excessive use of force. The week before the start of Brown's tenure, seven officers were dismissed and six suspended for harassing and beating black occupants of a local hotel. Mayor Kathy Whitmire indicated that she hired Brown because he was a "good manager" (Balz, 1982). However, some within the police force and city government were concerned about Brown's status as an outsider since he was not hired from within the department (Balz, 1982).

Brown's seven years in Houston were characterized as improving race relations and being in the vanguard in the use of community policing. He aimed to have each citizen know at least one patrol officer (Belkin, 1989a). As he had done in Atlanta, Brown set up police substations in order to facilitate neighborhood policing. He also instituted single-officer patrol cars in order to cover all of Houston (Belkin, 1989b; Blumenthal, 1989a). The force increased to 4,500 officers during his tenure (Kurtz, 1989). In addition, Brown increased the percentage of black police officers in the department to 14.1% from 8.5% (Belkin, 1989a). However, the union was often at odds with Brown, claiming that he lowered standards for recruitment, resulting in the hiring of irresponsible officers. In fact, when Brown left the department, three cases of police abuse overshadowed his departure. One involved the rape of a woman by an officer in a patrol car; that officer was sentenced to four months in prison. Another incident involved the arrest of an officer who was accused of telling several prostitutes he would release them from detention in exchange for sexual favors. A third incident involved the fatal shooting by police of a black motorist pulled over for speeding. However, according to Representative Craig Washington

(Democrat, Texas), "[Brown] was the most popular police chief ever," and that criticism of the police was "directed at the institution, not at him personally" (Kurtz, 1989, p. A3).

After Houston, Brown was appointed commissioner of the NYPD in 1990, where he served until 1992. According to Mayor David Dinkins, Brown was hired because of his reputation of improving police-community relations and race relations (Purdum, 1989). In New York City, according to media accounts, the top policing issue of the day was controlling drugs. Likewise, at the time Brown considered drugs to be the biggest problem facing the nation and believed making officers "a part of the community—not apart from the community" could have an impact on the crisis (Blumenthal, 1989b, p. B3). Brown stated:

> We can't just rely on the police. We've been doing what we've been doing since we've had America, relying on the police to address the crime problem, but it's not going to happen. We should stop responding to incidents and start solving problems. (Blumenthal, 1990a, p. B1)

In 1990 he instituted Operation Take-Back, which aimed to stem the tide of murders and robberies associated with the drug trade by paying 200 officers overtime pay each day to increase patrols, including foot patrols, in seven crime-prone neighborhoods. In addition, the Tactical Narcotics Team (TNT) was shifted to focus primarily on these seven high-risk areas (McKinley, 1990). Despite these initiatives, crime rates continued to rise and Brown was criticized. At a press conference on September 24, 1990, Brown defended his approach by saying that he could not change the fact that citizens allowed drug dealers to operate in their midst and that "the presumptions that cops are not doing their job" were frustrating (Blumenthal, 1990b, p. B1). He also stated:

> People know who the drug dealers are, but all too often we don't get the cooperation

that we need to address that as a major concern. We don't want people to do police work. We want them to be our eyes and ears. (Blumenthal, 1990b, p. B1)

Despite his public criticisms of citizen apathy, Brown was known for sharing his vision of community policing with citizens at a variety of gatherings. In his first year as commissioner, Brown attended hundreds of religious and civic meetings to explain citizen involvement in policing (James, 1991b).

In addition, budget freezes had required the police to do more with less (Blumenthal, 1989a). The force in 1990 numbered 26,300 officers, down by approximately 5,000 officers since 1974 (Blumenthal, 1990a). An initial hiring freeze accompanied Operation Take-Back (McKinley, 1990); however, by late 1990 a class of 500 entered the force, receiving six months of field training in community policing. Brown also hired civilian workers in order to relieve from desk jobs those officers who could be in the field (Purdum, 1990). Brown also believed that officers were "overspecialized" and that more officers needed to be deployed in a generalized community-policing capacity (Blumenthal, 1990b). In 1991, the department eliminated 14 specialized units, putting an additional 242 officers on patrol (McKinley, 1991).

As a nationally renowned police chief, Brown assumed a leadership role after the Rodney King beating in Los Angeles. Brown urged police chiefs across the country to focus on the problem of excessive use of force. He also held a "chiefs' summit" on April 16, 1991, in which chiefs from Atlanta, Baltimore, Boston, Bridgeport, Chicago, Dallas, St. Petersburg, San Diego, Tulsa, and other cities discussed the King incident and its ramifications (James, 1991a).

However, Brown's force was not immune from criticism. In the Crown Heights area of Brooklyn, riots broke out on August 19, 1991, after a Hasidic Jew, Yoseph Lisef, caused a fatal car accident, killing Gavin Cato, an African-American child. Subsequently, approximately 250 Crown Heights residents, mostly black youths, began rioting both against Hasidic Jews in the area and against police. In the first night of violence, bottles were thrown, a 17-year-old fired a gun at an officer, and a Yeshiva van was set on fire (McQuiston, 1991a). The next day, 12 police officers were injured during disturbances. In retaliation for Cato's death, Nelson Lemerick, Jr., an African-American man, stabbed Yankel Rosenbaum, a Hasidic student from Australia. During the second day of riots, rocks and debris were thrown at police, and police had to separate black and Jewish protesters ("Child's death," 1991). On the third day, several hundred black youths hurled rocks and bottles primarily against the police, who were accused of colluding with the Hasidim for failing to arrest Lisef (Kifner, 1991). According to the press, the focus of the riotous anger shifted from the accident to a generalized expression of anger at racism and the history of blacks in this country ("Three nights," 1991).

Brown seemed to take a backstage position to Mayor Dinkins during the crisis. One of his rare public appearances was at a public meeting on August 21, 1991, in which Brown explained to citizens why Lisef had not been arrested, although all witnesses agreed that he had run a red light prior to the accident. Citizens criticized Brown for quickly arresting Lemerick, a black man, while no arrest of Lisef was made after the accident. According to media reports, Brown's explanations did not quell a gasping crowd (Kifner, 1991; McQuiston, 1991b).[1] In a city report about Crown Heights released in 1993, Brown was characterized as making appropriate efforts to reach out to the community by galvanizing community leaders. However, he was criticized for failing to adequately direct field operations during the disturbances. Reportedly, he unnecessarily decentralized command, leaving decisions to field commanders. He also allegedly did not monitor ongoing developments or challenge and question field commanders ("Crown Heights report," 1993).

In a minor controversy in November 1991, Brown was criticized for dispatching two police arson specialists to Long Island to advise a friend of Mayor Dinkins about a fire on his property. Brown told reporters that he did not believe he acted unethically, although he admitted that it was not common to send out officers for such missions. In addition, he stated that the same service would be offered to the average citizen. Brown also emphasized that Mayor Dinkins's friend, though not residing in New York City, worked there (James, 1991c).

In 1992, Brown added corruption spot checks and surveillance to the arsenal of Internal Affairs procedures. Internal Affairs investigators, whose numbers were increased, began after-hours surveillance of officers going to restaurants, bars, and their homes, even targeting those officers not specifically suspected of corruption. In addition, sting operations stashed bags of cash in crime-prone areas to see if officers who discovered them vouchered all of the money. The program was instituted as a response to the uncovering of a cocaine ring operated by six police officers who were arrested in May 1992. Although Brown did not believe the corruption problem was systemic, he admitted that the drug trade could be tempting to officers because of the cash involved (Wolff, 1992a).

In a second riotous event during Brown's service as police commissioner, on July 6, 1992, a disturbance broke out in the Washington Heights section of Manhattan during a demonstration against the fatal police shooting of Jose Garcia, a Dominican immigrant. Groups of 50 to 100 people overturned garbage cans and lit fires; a police helicopter was hit by gunfire. There were 11 arrests for charges ranging from arson to disorderly conduct (Dao, 1992a). A man throwing bottles fell five stories and died after being pursued along rooftops by police (Dao 1992a; Rein, 1992). A second day of violence occurred when a police officer was shot in the foot; gunfire was heard and people threw jugs of water off rooftops (Rein, 1992). Riots ensued on a third day, and 2,000 police officers in riot gear descended on the neighborhood. On a sixth day, a firecracker injured two officers, including a captain; 14 people were arrested. In six days of violence, there were 139 arrests, 74 police officer injuries, 16 civilian injuries, and 121 car fires (Dao, 1992b).

As during the Crown Heights situation, Brown appeared to take a backseat to Mayor Dinkins in addressing the community and the press about the riots and was rarely mentioned in press accounts at the time. Yet Brown visited Washington Heights throughout the crisis, meeting with local community leaders. He indicated that he planned to institute community sensitivity training at the local precinct (Dao, 1992b). The Patrolmen's Benevolent Association criticized the mayor for not visiting officers in the Washington Heights precinct and offering them a morale boost, but they did not lodge a complaint against Brown's leadership (Finder, 1992).

Civilian complaints were up 19% in the first six months of 1992 compared to the same period the year before, according to the Civilian Complaint Review Board (CCRB). Brown told the press that it was due to the increase in new officers who tended to be "more aggressive." However, *The New York Times* reported that most complaints were lodged against officers who had been on the force for five and six years (Wolff, 1992b).

Brown's legacy in New York City extols his implementation of community policing, although only one precinct had completely adopted the model (Wolff, 1992c). Yet the commitment to the policy was deep; Mayor Dinkins said about community policing, upon Brown's resignation, "[I]t's no longer a program. This is how we do business" (Blumenthal, 1992, p. B4). Despite this success, the press characterized Brown as a reserved, not charismatic, leader (Wolff, 1992c).

While NYPD commissioner, Brown served also as the president of the International Association of Chiefs of Police from 1990 to 1991 (Blumenthal, 1989a). After leaving the job as NYPD commissioner, Brown returned to Houston to teach at Texas Southern University (Blumenthal, 1992) and directed the university Black Male Initiative program. From 1993 to 1996, he served on President Clinton's cabinet as the director of the White House Office of National Drug Control Policy ("Biography," n.d.). In 1998, he was elected as the first black mayor of Houston, serving until

January 2004. While mayor, Brown negotiated a pay package for the Houston Police Department, making it one of the highest-paid police forces in the nation. According to a journalist who interviewed city administrators, Brown's leadership was characterized as problematic in that he overly relied on his subordinates and was perceived as being unable to make decisions. He also was described as favoring a conservative bureaucratic method (Fleck, 2003).

The fact that Brown had "no love affair" with the unions can be traced to his tenure as chief of the Houston Police Department. Accused by the union of lowering the standards of recruitment and hiring irresponsible officers who were later involved in egregious cases of police misconduct, Brown left the police department stigmatized by a lack of integrity rather than celebrated for his innovations. The truth of the matter is that, as identified by the pentagon of police leadership model (in Chapters 1 and 2), recruitment and selection are of paramount importance for effective police leadership. Nevertheless, it is frequently the case that recruitment and selection are informed by political considerations and that the potential recruit who is sought by a given department represents more the "flavor of the month" requirements rather than an embodiment of the true qualifications for police work. The whole idea of identifying true competencies for effective policing, though hardly novel, was never really researched and investigated well enough to provide a standard to be modeled and replicated by police organizations. Finally, even if such competencies were fully identified, the dictates of reality end such idealistic notions. People who are attracted to the police profession are not always the ones who should be attracted to it, and recruiters often must compromise; frequently, the decision is based on who is available, not on who is desired. Regardless of these factors, the selection of some and not others always results in an out-group pointing an accusatory finger at the police chief, or any chief for that matter.

Brown's later appointment as police commissioner of the NYPD did not endear him to the local unions there either. His innovative approach to implementation of the ideas of community-oriented policing, and more specifically his priority of reducing police misconduct, was not something that the unions were willing to swallow easily. During the riots of Crown Heights and the second wave of riots in Washington Heights, Brown did show an overreliance on in-groups, whether within the organization itself or outside (by taking a backseat to Mayor Dinkins). Although Brown was not overtly criticized by the unions, which chose to criticize Mayor Dinkins instead, it was clear that Brown's preoccupation with the community members and their well-being, and his low profile as far as the police response to the riot was concerned, did not win much love from the local unions.

A police commissioner who attempts to clean up a police force on one hand and is overly preoccupied with the local community (to the point that the community is perceived as his in-group rather than the officers on the force) on the other will rarely gain real support from the unions. For better or worse, he will be perceived as an in-group/out-group leader.

After being mayor of Houston, Lee P. Brown worked as a visiting scholar at Rice University (D'Onofrio, 2009). In 2004, he opened his own consulting firm Brown Group International (BGI), becoming the company's President and CEO (D'Onofrio, 2009). In August 2005, Brown became a member of the Advisory Board of Stanford Group Company (CAMAC Energy Inc., 2011). In January 2006, Brown became the chairperson of the board of Unity National Bank (D'Onofrio, 2009). In January 2007, Brown's company BGI was hired to evaluate the New Orleans Police Department in the aftermath of Hurricane Katrina (D'Onofrio, 2009).

More recently, Brown was named Director of CAMAC Energy Inc. in April 2010 (CAMAC Energy Inc., 2011). He is also currently serving as a member of the Advisory Board at Swiftships Shipbuilders, L.L.C., and as a member of the Advisory Board of Carbon Motors Corporation (CAMAC Energy Inc., 2011).

[1] A grand jury subsequently decided not to bring charges against Lisef because they believed the crash was an accident and not a criminal act ("Justice," 1991).

CAREER HIGHLIGHTS

Chief Darrel Stephens
Newport News, Virginia; St. Petersburg, Florida; and Charlotte-Mecklenburg, North Carolina

Darrel Stephens earned a bachelor's degree in administration of justice from the University of Missouri at Kansas City and a master's degree in public administration from Central Missouri State University ("Chief Darrel W. Stephens," n.d.). Stephens began his policing career in the Kansas City, Missouri, Police Department. He was the assistant police chief in Lawrence, Kansas, from 1976 until 1980. From 1980 to 1983, he was the chief of police in Largo, Florida ("Chief Darrel W. Stephens," n.d.).

In 1983, Stephens was hired as chief of police in Newport News, Virginia, to confront corruption in the department (Rogers, 1999b). However, he is most well known for initiating community-oriented policing in the city. In the mid-1980s, Stephens recommended the demolition of the New Briarfield Apartments, a housing project that was notorious for drug use and that had the highest rate of burglaries in the city. Police interviews with residents revealed that people lived in fear of criminal activity and were concerned that physical deterioration of the buildings and vacant housing units attracted criminals. The police worked with members of other city agencies to pick up trash, dispose of abandoned cars, remove potholes, and sweep streets in order to reduce low-level crime before the project's eventual demolition. According to the Police Executive Research Forum (PERF), the burglary rate dropped 35% after these improvements (Wilson & Kelling, 1989).

From approximately 1986 to 1992, Stephens served as executive director of the PERF in Washington, D.C. ("Chief Darrel W. Stephens," n.d.). From 1992 until 1997, Stephens was chief of police in St. Petersburg, Florida. In assuming the job in St. Petersburg, Stephens inherited a 500-officer force in which infighting was common: blacks versus whites, union versus management, and managers versus civic leaders. In addition, police-community relations were tense. Stephens targeted community-oriented policing as a solution, which appeared to succeed in reaching out to the community but ultimately did not solve the divisions within the department (Roche, 1996a).

In pursuit of community-oriented policing, Stephens increased the visibility of police in the city, a move he believed would reduce citizens' fear of crime. This strategy involved creating three police substations in the city. Stephens also decentralized headquarters and permitted officers to police consistently in the same neighborhoods in order to build police–citizen partnerships. In addition, he proposed a program that would allow officers to take their patrol cars home. He hoped to spend $13 million on additional cars in order to assign them to specific officers. "I see the value of the cars to increase the visibility in the community and the neighborhoods. They will provide heightened sense of safety. My reason for proposing it has to do with increasing our visibility and decreasing fear, hopefully" (Roche, 1996a, p.1). However, the city council ultimately rejected the proposal (Roche, 1996a).

The most controversial event while Stephens served as chief occurred on October 24, 1996, when a white police officer, Jim Knight, killed an 18-year-old black motorist, Tyrone Lewis, the sixth victim of police shootings that year. Lewis had refused to roll down his tinted windows or follow any commands at a traffic stop for speeding. According to police accounts, the car lurched forward and nearly hit Officer Knight, who fired shots at Lewis. Minutes after the shooting, crowds took to the streets, burning buildings and throwing rocks and bottles in a 25-block area. Additionally, a patrol car was firebombed, two media vans were burned, and fires were set at a police substation and a post office; 11 people were injured, including a police officer and a photojournalist who were beaten. A total of 28 buildings were burned and 20 people were arrested (Flores, 1996).

Stephens responded by defending Officer Knight's conduct in the shooting of Lewis but placed Knight on administrative leave pending an investigation. City council member Ernest Filayu told the press that the frustrations exhibited by the riots went "deeper than a shooting" and that police come into black neighborhoods "with their guns cocked and their attitudes cocked. We have a legal harassment system" (Williams, 1996, p. A3). On October 27, 1996, 500 citizens attended a school desegregation meeting called by community leaders, but it quickly became an outpouring of frustration at the police department about the Lewis killing and the rioting. At another meeting that night, called by the Community Alliance, Stephens and the mayor attended and took notes. Many called on them to get to the bottom of the issue of police brutality that sparked the riot. Stephens spent the Monday after the riots praising his officers for the job they did during the disturbance. The Police Benevolent Association, however, criticized the department for not having enough portable radios and riot gear. Stephens later acknowledged that equipment arrived late on the scene (Landry, 1996).

A second disturbance in St. Petersburg occurred after a grand jury cleared another officer involved in the shooting of Lewis, rejecting the findings of a civilian review board that had found that Officer Knight had violated department policy in responding to the incident. Stephens responded in a memorandum to his officers stating that the policy allows officers to shoot to protect themselves if they have done everything in their power to get out of harm's way. He indicated that although Lewis's car was slow-moving, Knight was bumped by it several times and could not get out of the way (Howard et al., 1996). During the unrest, there were 78 reported fires and an armed robbery. A drive-by shooting injured one person; three additional people were injured in another shooting. An officer was shot but not seriously injured (Thompson, 1996a). There were 21 burglaries and 6 thefts and assaults (Roche, 1996c).

The disturbance geographically centered in the blocks near the house of the black militant group, National People's Democratic Uhuru Movement, where youths roamed during the eight-hour disturbance. The Uhuru philosophy is one of violent resistance against a judicial system the Uhuru view as fraudulent (Thompson, 1996a). Earlier in the day, police arrested three Uhuru members on outstanding charges, reportedly in order to prevent these militants from stirring up trouble during any unrest that was likely to follow the grand jury verdict. In one demonstration prior to the second riot, the Uhurus held a mock tribunal and ordered the execution of Stephens and the mayor, Dave Fisher, by electric chair because they were allegedly guilty of genocide (Rogers, 1999a). Later in the evening, police pulled up in front of the headquarters and were fired on by Uhurus and supporters gathered around the station house. The police used tear gas to dispel the crowd, causing a stampede. In the following unrest, an officer and a citizen were shot, a motorist's jaw was broken by an object hurled into his window, and 33 fires were caused by Molotov cocktails. Near the headquarters, a man opened fire with an automatic on four police officers. The police fired back; there were no injuries in the exchange (Thompson, 1996a).

Black church and civic leaders criticized the arrest of the Uhuru leaders as being heavy-handed and complained that it exacerbated already troubled police-minority relations. Stephens indicated, however, that he had reasons for believing that the Uhurus were planning to react

violently if Officer Knight was not charged and so used the preventive strategy of arresting some of them who had outstanding warrants (Roche, 1996b). Police intelligence reports indicated that the Uhurus had stockpiled bricks and bottles in neighborhood trash cans to use in a riot if the grand jury decided in favor of Officer Knight (Thompson, 1996a). Furthermore, complaints about the use of tear gas were also made. Stephens defended the actions of the police officers against the tear gas complaints, saying they were following his strategy of isolating the unrest and holding instigators accountable. He also said that the demonstrators were given a five-minute warning to disperse before the tear gas was set off. Yet he also revealed that the specific decision to use tear gas was made by police supervisors in the field. Police union leaders criticized Stephens for not having provided officers with extensive riot training. Stephens responded that there had not been a major disturbance in the area for two decades before the riots sparked by the killing of Lewis (Roche, 1996b).

After the riots and criticism, Stephens admitted publicly in November 1996 that he was not sure he would survive as chief. He stated, "Any police chief that encounters a situation like we've encountered, people ought to be asking 'Is this guy doing the sorts of things he should have done the last several years?' " (Thompson, 1996b, p. 1). However, he also highlighted his department's accomplishments, including decreasing crime and increasing a sense of security among citizens. He reiterated his commitment to community policing despite the unrest, saying that approximately 20% of the black community supported the concept. He also stated that everyone in the city would need to address the socioeconomic conditions and racial tensions that act as barriers to effective policing (Thompson, 1996b).

A month later, the police union released survey results indicating that out of 307 officers who responded, 76% viewed Stephens as weak and ineffective. They called for an investigation into the chief's actions during the riots. Stephens responded by welcoming the outside review, indicating that his job was not about meeting the expectations of his officers but serving the interests of the community. Union President Jack Soule underscored the criticism, "[Stephens] is a sensitive person when it comes to dealing with issues in the community. Where we feel he has failed is with his building a foundation of a partnership with the employees" (Roche, 1996d, p. 1B). A subsequent State Advisory Board to the U.S. Commission on Civil Rights investigated the police response to the riots. Mary Frances Berry, the commission chairperson, questioned Stephens for not personally leading the officers who were to arrest the Uhuru leaders on outstanding charges, suggesting that the tactic had inflamed the black community and been counterproductive. Stephens later acknowledged that the arrest of the Uhurus was a mistake in judgment (Smith, 1996). Despite the riots and subsequent controversy, press reports indicate that the black community generally credited Stephens with trying to calm racial tension (Rogers, 1999a).

Throughout his tenure as chief, Stephens struggled with low officer morale. A 1995 survey of 58% of officers in the Police Benevolent Association indicated that 93% of them felt officer morale was low or very low. More than 50% criticized police managers and Stephens regarding safety issues, retirement plans, and promotions. Between 1993 and 1995, four reverse discrimination suits were filed with the U.S. Equal Opportunity Commission. In response to the survey, Stephens told reporters that he was unsure what to do about the morale problem but that he would look into it (Thompson, 1995).

In June 1997, the St. Petersburg city administrator resigned and the mayor offered the position to Stephens. Struggling with the police department after the riots, Stephens indicated that he took the job at a "soul-searching" moment in his career (Ryan, 1998, p. 1B). The mayor did not expressly remove Stephens because of the racial tension and riots, but this was the public perception of his departure (Thompson, 2001). Stephens remained as city administrator until September 1999 when he left to become the chief of the Charlotte-Mecklenburg Police Department (Ryan, 1998).

Charlotte had a force of 1,500 sworn officers. In continuing his efforts at community-oriented policing, Stephens maintained advanced police technology and equipment in the department (Stephens, n.d.). He also focused on other city functions, such as swimming pool and park hours, and analyzed how these might affect the crime rates. Graffiti elimination and landscaping were also employed to reduce low-level street crimes (Rogers, 1999b).

In addition, Stephens formed a cold case team in February 2003 to investigate over 360 unsolved homicides spanning 40 years. Homicide detectives admitted that they rarely had time to look at old files. The team consisted of two experienced detectives, an FBI agent, two retired NYPD detectives, and a criminal justice professor. They were to review the cases first to find out which were the most solvable, with particular focus on which cases might have DNA evidence that could be analyzed using new technology ("Police designate," 2003).

One controversy Stephens confronted as chief in Charlotte began on May 8, 2002, after a bungled sting operation meant to capture a man who had previously assaulted and kidnapped his girlfriend. The girlfriend, a 33-year-old woman, was used as "bait" but was missing for 16 hours after the sting operation failed and she had left with the suspect without the officers' knowledge. Stephens immediately admitted that the sting operation was a mistake and that the detective who devised the plan, although he was well-meaning, had used the wrong methods to apprehend the suspect. Stephens also mentioned, however, that there was no department policy against using victims in sting operations (Whitmore, 2002).

Stephens was one of the main pioneers/proponents of community-oriented policing. It is likely that it was that approach which alienated local police unions from being an ideal in-group category. He was accused by the Police Benevolent Association of not providing enough radios and riot gear after the Lewis shooting; 76% of respondents to a union-initiated survey identified him as a weak and ineffective leader. Finally there came a point-blank accusation by the union president who blamed Stephens for failing to build a partnership with his employees at the expense of building a stronger and more sensitive partnership with the local community. This last accusation was, practically speaking, an affirmation of the in-out group concept, when the larger community becomes an in-group for the police chief and the prevailing perception among the union leaders is that, as a result, the officers become the out-group.

The LMX theory of leadership's recommendation of creating an in-group of all is unrealizable in a democratic policing environment, in which there is a competing pull on police leaders to be accountable both to their own officers and to the community at large. One interesting parallel between these two chiefs is their endorsement of community-oriented policing. Yet their careers were both stigmatized by accusations that they favored groups outside the police organization. Their experience raises the question of whether full commitment to or an orientation toward community-oriented policing makes leaders prone to accusations of in-group/out-group polarity. Does it increase the chances that a police department's union will perceive that officers have become scapegoats, one large out-group for the police chief in charge, whose in-group has become the external community?

In 2006, Stephens received an Honorary Doctorate of Laws Degree from Central Missouri State University (Johns Hopkins University School of Education, 2010). Two years later, in March 1, 2008, he retired from the Charlotte-Mecklenburg Police Department and became an instructor in the Public Safety Leadership Program at Johns Hopkins

University in June 2008 (Johns Hopkins University School of Education, 2010; PoliceLink, 2011). That same month, he was appointed Director of State and Local Programs at Johns Hopkins University (LinkedIn, 2011).

In 2008, he established the consulting firm Darrel Stephens Group LLC (LinkedIn, 2011). Stephens also received four awards that year: the United States Secret Service Director's Honor Award, the U.S. Drug Enforcement Agency Outstanding Contributions to Drug Law Enforcement Award, the U.S. Bureau of Alcohol, Tobacco, and Firearms, Charlotte Division, Outstanding Leadership Award, and the Major Cities Chiefs Association Outstanding Leadership Award (LinkedIn, 2011).

Stephens was inducted into the Evidence Based Policing Hall of Fame May 2010 (Center for Evidence Based Crime Policy, 2010). He also became the Executive Director of the Major Cities Chiefs Association in October 2010 (LinkedIn, 2011).

Exercise

The police chief and commissioner detailed in this chapter have been examined through the prism of leader–member exchange theory. Using what you have learned from other theories in this book or from the summaries (as indicated in the Introduction) you have read thus far, it is possible to consider their approaches through the lens of different theories. What other theories can you find applicable to these leaders? Are there lessons to be learned from those applications, or do those theories indicate alternative paths or approaches that might have yielded a more successful outcome?

References

Balz, D. (1982, March 10). Atlanta official hired by Houston as police chief. *The Washington Post*, p. A24.
_____ & A. Harris (1982, April 20). Atlanta's troubleshooter is installed as Houston chief to clean up police. *The Washington Post*, p. A7.

Belkin, L. (1989a, December 19). Man in the news: A chief known for turning strife into calm. *The New York Times*, p. B9.
_____ (1989b, December 21). For new police commissioner, towering challenges: Stranger to New York,

Brown says first act will be to "reach out." *The New York Times*, p. B1.

Biography of Lee P. Brown (n.d.). Office of National Drug Control Policy. Retrieved August 7, 2003, from http://clinton1.nara.gov/White_House/EOP/ondcp/html/Lee_Brown.html.

Blumenthal, R. (1989a, December 21). For new police commissioner, towering challenges: Rampant drug abuse and record murder toll compete with 911 calls and budget freezes. *The New York Times*, p. B1.

_____ (1989b, December 22). Police dept. is praised by next commissioner. *The New York Times*, p. B3.

_____ (1990a, August 3). Crime and commissioner: Pressure mounts for results. *The New York Times*, p. B1.

_____ (1990b, September 25). Head of police assails media on job image. *The New York Times*, p. B1.

_____ (1992, August 6). Brown says community policing will endure. *The New York Times*, p. B4.

CAMAC Energy Inc. (2011, May 29). *Bloomberg Businessweek*. Retrieved May 29, 2011, from http://investing.businessweek.com/research/stocks/people/person.asp?personId=23722299&ticker=CAK:US&previousCapId=62604221&previousTitle=CAMAC%20ENERGY%20INC.

Center for Evidence Based Crime Policy. (2010). *Darrel W. Stephens*. Retrieved from George Mason University Website: http://gunston.gmu.edu/cebcp/HallofFame/Stephens.html.

Chief Darrel W. Stephens (n.d.). Retrieved July 17, 2003, from http://www.cicp.org/Boardmembers.htm.

Child's death sparks outrage, riot, slaying (1991, August 21). *The Record*, p. A3.

Crown Heights report: From the report on police actions and Commissioner Brown (1993, July 21). *The New York Times*, p. B7.

Dao, J. (1992a, July 7). Angered by police killing, a neighborhood erupts. *The New York Times*, p. A3.

_____ (1992b, July 10). Washington Heights calm shattered by disruptions. *The New York Times*, p. B3.

Dinoeh, R.M. & R.C. Liden (1986). Leader–member exchange model of leadership: A critique and further development. *Academy of Management Review*, 11, 618–634.

D'Onofrio, A. (2009, June 24). Today in Texas history: Lee Brown becomes Houston's first African American mayor. *Houston Chronicle*. Retrieved May 29, 2011, from http://blog.chron.com/txpoto mac/2010/01/today-in-texas-history-lee-brown-becomes-houstons-first-african-american-mayor/.

Duchon, D., S.G. Green & T.D. Taber (1986). Vertical dyad linkage: A longitudinal assessment of antecedents, measures, and consequences. *Journal of Applied Psychology*, 71, 56–60.

Dunegan, K.J., D. Duchon & M. Uhl-Bien (1992). Examining the link between leader–member exchange and subordinate performance: The role of task analyzability and variety as moderators. *Journal of Management*, 18 (1), 59–76.

Erdogan, B., M. Kraimer & R.C. Liden (2002). Person-organization fit and work attitudes: The moderating role of leader–member exchange. *Academy of Management Proceedings* F1–F6.

Finder, A. (1992, July 9). Tension in Washington Heights: Dinkins, amid crowd, nurtures fragile peace. *The New York Times*, p. A1.

Fleck, T. (2003, May 15). Parting shots: Foes of Mayor Brown want to see him gone. And a lot of his friends feel that way, too. *Houston Press*, News Section. Retrieved November 3, 2004, from http://www.houstonpress.com/issues/2003-05-15/news/feature.html.

Flores, I. (1996). St. Petersburg police restrict gun, gas sales day after rioting. *Chattanooga Free Press*, p. A1.

Graen, G. (1983, May–June). Bias in management research? A defense. *Business Horizons*, 42–45.

_____ & J.F. Cashman (1975). A role making model of leadership in formal organizations: A developmental approach. In J.L. Hunt & L.L. Larson (Eds.), *Leadership frontiers* (pp. 143–165). Kent, OH: Kent State University Press.

_____ & M. Uhl-Bien (1991). The transformation of professionals into self-managing and partially self-designing contributions: Toward a theory of leader-making. *Journal of Management Systems*, 3 (3), 33–48.

Harris, A. (1982, March 2). Atlanta ends special probe into killings. *The Washington Post*, p. A1.

Howard, P.E., S. Thompson & D. Sommer (1996, November 14). Jury report sparks unrest. *The Tampa Tribune*, p. 1.

James, G. (1991a, April 9). Brown sponsors a police summit on excess force. *The New York Times*, p. B2.

_____ (1991b, August 14). Brown's gospel: Community policing. *The New York Times*, p. B1.

_____ (1991c, November 14). Brown defends sending officers to Dinkins's L.I. friend. *The New York Times*, p. B3.

Johns Hopkins University School of Education. (2010). *Stephens Darrel*. Retrieved from Johns Hopkins University School of Education Website: http://education.jhu.edu/faculty/faculty_and_staff/855486BEAED3F201D3597F98F47840DB/Darrel_Stephens.

Justice in Crown Heights (1991, September 7). *The New York Times*, p. 22.

Kifner, J. (1991, August 22). Tension in Brooklyn: Clashes persist in Crown Heights for 3d night. *The New York Times*, p. B1.

Kozlowski, S.W.J. & M.L. Doherty (1989). Integration of climate and leadership: A neglected issue. *Journal of Applied Psychology*, 6, 546–553.

Kurtz, H. (1989, December 19). Texan to head N.Y. police: Brown's term marred by 2 recent killings. *The Washington Post*, p. A3.

Landry, S. (1996, October 29). Anger comes out at forum. *St. Petersburg Times*, p. 1A.

LeBlanc, P.M., R.D. deJong, J. Geersing, J. Furda & I.H. Komproe (1993). Leader member exchanges: Distinction between two factors. *European Work and Organizational Psychologist*, 3 (4), 297–309.

Liden, R.C. & G. Graen (1980). Generalizability of the vertical dyad linkage model of leadership. *Academy of Management Journal*, 23 (3), 451–465.

LinkedIn. (2011). *Darrel Stephens*. Retrieved from LinkedIn Website: http://www.linkedin.com/pub/darrel-stephens/10/1b3/b04.

Maurer, T.J., H.R. Pierce & L.M. Shore (2002). Perceived beneficiary of employee development activity: A three-dimensional social exchange model. *Academy of Management Review*, 27 (3) 432–444.

McKinley, J.C. (1990, July 19). Extra patrols to be added for 7 precincts. *The New York Times*, p. B1.

_____ (1991, February 9). Anti-crime plan will curb violence, commissioner says. *The New York Times*, p. 27.

McQuiston, J.T. (1991a, August 20). Fatal crash starts melee with police in Brooklyn. *The New York Times*, p. B1.

_____ (1991b, August 22). Tension in Brooklyn: Case weighed in car accident that killed boy. *The New York Times*, p. B3.

Norris, W.R. & R.P. Vecchio (1992). Situational leadership theory: A replication. *Group and Organizational Management*, 17, 331–342.

Northouse, P.G. (2004). *Leadership: Theory and practice* (3rd ed.). Thousand Oaks, CA: Sage.

Police designate investigators for unsolved killings (2003, February 11). *The Associated Press,* state and local wire.

PoliceLink. (2011). *Chief of police Darrel Stephens Charlotte-Mecklenburg, NC police department*. Retrieved from PoliceLink Website: http://policelink.monster.com/albums/143237-cheif-of-police-darrel-stephens-charlotte-mecklenburg-nc-police-department.

Purdum, T.S. (1989, December 19). Dinkins names Houston's chief to be his police commissioner. *The New York Times*, p. A1.

_____ (1990, September 19). Police chief wants to add more civilians. *The New York Times*, p. B1.

Rein, L. (1992, July 8). Violence flares anew: Cop is shot in fallout from killing. Unrest in Washington Heights. *The Record*, p. A1.

Roche, T. (1996a, March 31). Chief wants to make officers more visible. *St. Petersburg Times*, p. 1.

_____ (1996b, November 15). Police had not planned for immediate violence. *St. Petersburg Times*, p. 1A.

_____ (1996c, November 20). Unrest adds up to large expense. *St. Petersburg Times*, p. 1B.

_____ (1996d, December 4). Union president calls for police chief's ouster. *St. Petersburg Times*, p. 1B.

Rogers, D.K. (1999a, August 26). Stephens to head Charlotte force. *St. Petersburg Times*, p. 1B.

_____ (1999b, September 19). Stephens anxious to begin new post. *St. Petersburg Times*, p. 1.

Ryan, K. (1998, June 14). City administrator Darrel Stephens: From top cop to city bureaucrat. *St. Petersburg Times*, p. 1B.

Scandura, T., G.B. Graen & M.A. Novak (1986). When managers decide not to decide autocratically: An investigation of leader–member exchange and decision influence. *Journal of Applied Psychology*, 71, 579–584.

Schriesheim, C.A., S.L. Castro, X. Zhou & F.J. Yammarino (2001). The folly of theorizing "A" but testing "B": A selective level-of-analysis review of the field and a detailed leader–member exchange illustration. *Leadership Quarterly*, 12, 515–551.

Smith, A. (1996, December 5). Rights panel grills mayor, police chief. *St. Petersburg Times*, p. 1A.

Steiner, D.D. (1988). Value perceptions in leader–member exchange. *Journal of Social Psychology*, 128 (5), 611–618.

Stephens, D. (n.d.). Message from the Charlotte-Mecklenburg police chief. Retrieved on July 17, 2003, from http://www.charmeck.nc.us/Departments/Police/About+Us/Command+Staff/home.htm.

Thompson, S. (1995, August 11). Union survey finds cop morale low. *The Tampa Tribune*, p. 1.

_____ (1996a, November 15). Uhuru black militants thrust into spotlight with new unrest. *The Tampa Tribune*, p. 12.

_____ (1996b, November 27). Stephens admits his job could be in jeopardy. *The Tampa Tribune*, p. 1.

_____ (2001, December 23). Tenuous race relations haunt police chiefs. *The Tampa Tribune*, p. 1.

Three nights, three moods in the streets (1991, August 25). *The New York Times*, p. 37.

Vecchio, R. & B. Gobdel (1984). The vertical dyad linkage model of leadership: Problems and prospects. *Organizational Behavior and Human Performance*, 34, 5–20.

Whitmore, T. (2002, May 8). Police chief: Sting operation a mistake. *The Associated Press*, state and local wire.

Williams, M. (1996, October 27). St. Petersburg searches its soul: Riot highlights racial tension belying progress. *The Atlanta Journal and Constitution,* A3.

Wilson, J.Q. & G.L. Kelling (1989, February). Making neighborhoods safe. *Atlantic Monthly,* 46–52.

Wolff, C. (1992a, July 1). Brown to add inquiry power in police graft. *The New York Times,* p. B1.

_____ (1992b, July 30). Police review board says public complaints are up. *The New York Times,* p. B2.

_____ (1992c, August 4). Brown's resignation. Brown legacy: Community policing. *The New York Times,* p. B2.

Yammarino, F.J., F. Dansereau & C.J. Kennedy (2001). A multiple-level multidimensional approach to leadership: Viewing leadership through an elephant's eye. *Organizational Dynamics,* 29 (3), 149–163.

Yrle, A.C., S.J. Hartman & W.P. Galle (2002). An investigation of relationships between communication style and leader–member exchange. *Journal of Communication Management,* 6 (3), 257–268.

_____, S.J. Hartman & W.P. Galle, Jr. (2003). Examining communication style and leader–member exchange: Considerations and concern for managers. *International Journal of Management,* 20 (1), 92–100.

When the Chief Becomes the Force
Transformational Theory

OVERVIEW OF THE THEORY

In transformational leadership, the personality of the leader stimulates change by raising consciousness, motivation, and morale. Here it is illustrated through the examples of Chief Reuben Greenberg of the Charleston, South Carolina, Police Department and Chief Daryl Gates of the LAPD. Although both had careers with many facets, the key for both was the degree to which their personalities succeeded in creating desired change.

This theory was first articulated by Burns (1978), who identified a process by which leaders affect the attitudes and motivations of followers. According to Burns (1978), transformational leadership involves leaders tailoring their leadership to the needs of their followers to move them into a state of elevated goals and performance; transcending the implicit or explicit demands of the task. Burns (1978) refers to two developmental continua: follower motivation and moral development. The former involves inspiring followers to meet self-actualization needs and to perform at their highest potential and the latter, to the leader's ability to be inspirational and uplifting, moving followers to a higher moral stage by urging followers to transcend their own self-interest for the sake of the organization (Bass, 1985). According to Bass and Avolio (1994), transformational leadership can range in style from democratic and participative to directive and authoritarian.

Burns (1978) contrasts transformational leadership with transactional leadership, characterized as exchanges between leaders and followers that remain within the limits of self-interest. Under this model, followers are motivated by what they will gain from the task (e.g., salary, bonuses), and leaders are satisfied with the mere fulfillment of contractual obligations (the *contingent reward*). Transactional leadership can also take the form of *laissez-faire management*. Here, leaders are passive and avoid taking action. Thirdly, transactional leadership may also be *management-by-exception* in which the leader monitors followers, intervening only if a corrective action is required (Bass, 1999).

The transformational/transactional distinction represents the difference between mere management and true leadership. Burns (1978) originally conceptualized the two as opposite ends along a spectrum of leadership; however, Bass (1985) contends that the two are separate concepts, and that a leader can be both transactional and transformational. Regardless, the distinction between management and leadership is rooted in historical trends. In the 1970s and 1980s, American companies were unable to cope with changing

Ethics and Leadership

- Leaders are to encourage followers to develop a set of values that emphasizes justice, liberty, and equality.
- These values give transformational leadership its moral purpose.
- Transactional leadership lacks a moral center because it is inherently uncritical of the people involved in the leadership exchange.

(Burns, 1978)

economic demands; a problem that researchers blamed on too much management and too little leadership and managers lack of skill/talent (Bennis & Nanus, 1985; Conger, 1999).[1]

Transformational leadership consists of four factors: idealized influence (charisma), inspirational leadership, intellectual stimulation, and individualized consideration. *Idealized influence* (charisma) and *inspirational leadership* occur when followers want to identify with leaders because of their ability to define goals, to state how they can be accomplished, to set high standards, and to exhibit confidence and determination. *Intellectual stimulation* involves tapping into followers' creativity. *Individualized consideration* refers to a coaching style of leadership where leaders concern themselves with the career development of their followers (Bass, 1999; Bass & Avolio, 1994).

IMPORTANCE OF CHARISMA IN LEADERS

Max Weber (1947) conceptualizes charismatic leadership as the followers' belief that their leader has extraordinary qualities. House (1976) identified indicators of charismatic leadership, including follower perceptions and leader traits and behaviors. Charismatic leaders are self-confident and seek power. They are able to manage their image, articulate visions, communicate high expectations, and instill confidence in followers (De Vries et al., 1999). House's (1976) model focused on the dyadic relationship between leaders and followers. Criticism of House's (1976) conception

Ethics and Transactional Leadership

Giampetro-Meyer et al. (1998)

- Transactional leadership can lead to ethically questionable behavior on the part of leaders.
- Due to "craft ethics," leaders do not reflect on what they personally think is right but merely follow what those on the job or in the workplace believe is ethical.

Turner et al. (2002)

- Leaders who use the transactional style of leadership do not become more effective regardless of whether they have the capacity to reason morally.
- This is because, by definition, transactional leadership is merely a dyadic exchange to satisfy individual needs.

---

[1] See my review of trait theory (Chapter 12) for a synopsis of Bennis and Nanus (1985). According to their conception of transformational leadership, transforming leaders have vision, act as social architects, create trust, and use creative deployment of self through positive self-regard. Although Bennis and Nanus believe that these strategies of leadership are skills that can be learned, the skills appear to also operate much like traits in that they describe what characteristics a transforming leader should possess.

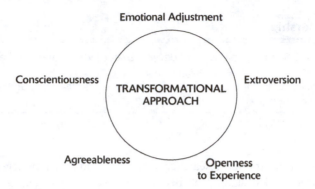

FIGURE 5-1 Five-Factor Model of Personality

Source: Adapted from Judge and Bono, 2000.

include that charismatic leaders are too varied to identify specific traits (Bryman, 1992; De Vries et al., 1999). In subsequent theory, House and Shamir (1993) moved to the collective, proposing that charismatic leadership occurs when leaders link their followers' sense of identity to the larger goals of the organization; thereby leading to the internalization of those goals.

The relationship between personality and transformational leadership was tested by Judge and Bono (2000) using the five-factor model of personality, providing empirical support for the centrality of charisma in the transformational model (see Figure 5-1). The five factors are extroversion, agreeableness, conscientiousness, emotional adjustment, and openness to experience, all of which are theoretically positively related to the likelihood that a leader will be transformational. According to Judge and Bono (2000), agreeableness was the strongest predictor of transformational leadership behaviors, particularly that of charisma. Extroversion and openness to experience were also positively related. Conscientiousness was not correlated with transformational leadership. Emotional adjustment was operationalized using measures of its opposite, neuroticism. There was no correlation between neuroticism and the absence of transformational leadership. Although as a whole the five-factor model was not an ideal predictor of transformational leadership behavior, this study's importance rests on the empirical finding that charisma as a measure of agreeableness is the best predictor of transformational behavior, as suggested by the approach of House (1976) and those who follow his paradigm.

Others have hypothesized that charisma is a behavioral phenomenon (Conger & Kanungo, 1998). Here, charisma is based on followers' perceptions of leader behavior. Leaders evaluate the talents and needs of subordinates and then expose the weaknesses of the status quo within the organization, followed by the formulation of new goals. Conger and Kanungo (1998) characterize charismatic leadership as leading toward radical change. This is similar to the Tichy and DeVanna (1990) approach, which emphasizes that recognizing the need for change is a key characteristic of transformational leaders. They also propose that the charismatic vision is based on the circumstances of the external environment. Likewise, leaders may articulate new processes and goals that are necessary due to pressures mounting from outside the organization (Conger, 1999).

Need for Charismatic Leadership

In a follower-centered approach, De Vries et al. (1999) tested the notion that followers differ in their need for charismatic leadership. They define the need for leadership as the extent to which followers desire the leader to outline the path toward individual, group, or organizational goals.

Charismatic Leader's Effect on Subordinates

Subordinates are more dependent rather than empowered when under a charismatic leader (De Vries et al., 1999).

or

Charismatic leaders empower their followers to greater independence and career development (Bass & Avolio, 1994).

Subjective needs are those associated with individual goals; *objective needs* characterize group and organizational goals. De Vries et al. (1999) found that subordinates who reported having charismatic bosses also exhibited a higher need for leadership. Need for leadership acted as a moderator variable between level of leaders' charisma and measures of performance. Likewise, subordinates are more dependent rather than empowered when they are under a charismatic leader; contradicting traditional transformational theory (Bass & Avolio, 1994).

META-ANALYTIC STUDIES ON TRANSFORMATIONAL LEADERSHIP

Lowe et al. (1996) found that transformational leadership behaviors are correlated with subjective and objective measures of leadership effectiveness and that this can be generalized across types of organizations and levels of management. Furthermore, individualized consideration, intellectual stimulation, and idealized influence (charisma) not only are correlated with effectiveness but also are strongly correlated with each other (Lowe et al., 1996). Fuller et al. (1996) also found that transformational leadership is related to leader effectiveness.

TRANSFORMATIONAL LEADERSHIP IN MILITARY CONTEXTS

Several studies testing transformational leadership using samples of military leaders and subordinates have been undertaken. Transformational leadership, operationalized using the Management Leadership Questionnaire (MLQ instrument) developed by Bass and Avolio that tests the four factors of this style of leadership (individualized consideration, intellectual stimulation, inspirational leadership, and idealized influence), was empirically related to subjective and objective performance evaluations in a sample of U.S. Navy leaders and followers. Measures of transactional leadership were not found to be significantly related to performance (Yammarino et al., 1993).

Transformational Leaders and Moral Reasoning

- Transformational leaders transcend self-interest goals to a more mature set of values, culminating in a sense of universalistic ethical principles of fairness, equality, and justice. These values form ends in themselves for mature leaders.
- Moral reasoning has been empirically connected to leadership effectiveness, using a measure of moral reasoning and subordinate ratings of leaders in a sample of managers.
- Moral reasoning is considered a cognitive function that can develop over a life span.
- Leaders with the cognitive capacity to reason morally will use this higher degree of transformational leadership and will be more likely to behave in a way that serves the collective good.

(Kuhnert and Lewis, 1987)

Masi (2000) studied transformational leadership among recruiting units of the U.S. Army. What Masi (2000) found is that there is a significant relationship between transformational leadership behaviors and motivation of subordinates; however, there is a negative relationship between transformational leadership behaviors and commitment to quality and productivity. Masi (2000) likewise concluded that in structured settings, leaders have only a marginal impact on the norms of their units, suggesting that structural changes (job redesign, revised human resources practices) would be needed to effect change in commitment to quality and productivity.

When Shamir et al. (1998) tested the House and Shamir (1993; see above) conception of charisma and self-identity, transformational theory was not supported by their findings. The only factor relating to measures of subordinate trust of the leader, identification with the leader, increased motivation, and willingness to sacrifice themselves for their military unit (measures of the success or failure of the charismatic leadership) was emphasis on collective identity. Charismatic traits (ideological approach, high devotion, high motivation, and high role involvement) on the part of leaders were cynically interpreted by subordinates as reflecting the interests of the military at large rather than the subordinates' own interests.

Dvir et al. (2002) conducted a field experiment involving 54 military leaders within the Israeli Defense Forces, their 90 direct followers (immediate subordinates; e.g., for a lieutenant, his/her sergeant), and 724 indirect followers (e.g., for a lieutenant, the officers commanded by his/her directly subordinated sergeant). The experiment subjected seven squads of platoon leaders and cadets to transformational leadership training. Five squads were assigned to the control group and received no transformational leadership training. In this study, a critical-independent approach, extra effort, and collectivist orientation were the dependent variables, all of which increased with the experimental group over time and decreased with the control group, regarding both direct and indirect followers, thus supporting the efficacy of transformational leadership. However, transformational leadership was not significantly related to direct followers' active engagement, internalization of moral values, and self-actualization needs. Regarding indirect followers, results suggested that transformational leadership creates a stronger relationship between direct and indirect followers, thereby improving the performance of indirect followers.

Dvir and Shamir (2003) continued studying the Israeli Defense Forces by testing whether followers' developmental characteristics predicted transformational leadership. The sample was the same as for the above study. However, in this study the organization's moral values, collectivist orientation, critical-independent approach, active task engagement, and self-efficacy were independent variables, measured using various survey instruments. The researchers hypothesized that the independent variables positively predicted transformational leadership over time. Results showed that the hypothesis was supported among indirect followers. However, direct followers showed a negative relationship. The researchers suggested three mechanisms to explain the unexpected outcome regarding direct followers. According to a threat explanation, the empowerment of direct followers acts as a threat to the power of their leaders; therefore, direct followers may suppress their level of development. The indirect followers, however, pose no threat to their leader's status, so their initial development level can be high. A compensatory explanation involves leaders perceiving direct followers as weak and thereby asserting their transformational leadership to a greater extent. This relationship does not exist between the leader and indirect followers because indirect followers do not share leadership responsibilities as leaders as direct followers do. A third explanation involves the potential disillusionment of direct followers who have daily face-to-face interactions with their leader.

TRANSFORMATIONAL AND TRANSACTIONAL LEADERSHIP STYLES AND THE CASES OF CHIEF REUBEN GREENBERG AND CHIEF DARYL GATES

Scholars define transformational leadership as an approach that differentiates between management and true leadership. It romanticizes the notion of a leader as someone who fully transforms the organization he/she heads to the extent that the agency is identified more with the leader's personality and his/her impact on subordinates and the organization than with the actual changes he/she creates and implements. Leadership techniques and philosophies and their results are emphasized less than the one individual who represents the culture of a given police organization. It frequently goes even a step further: The transformational leader *is* the organization that he/she leads.

The notion of a transformational leader portrayed by Kuhnert and Lewis (1987)—who depict transformational leaders as those who transcend self-interest goals to a more mature set of values that culminates in a sense of universalistic ethical principles of fairness, equity, and justice—does not necessarily fit the environment of policing. It is hard to envision the truly altruistic police chief who elevates himself/herself above and beyond self-interest goals. One wants to transform a police organization based on one's personal vision of what is right; this personal vision does not necessarily transcend self-interest goals. After all, if one believes that a certain vision should be implemented, it is based on the fact that this particular vision appears to be what projects one's values and beliefs, and it is self-serving, by default, when one implements that dream.

On the other end of the spectrum is transactional leadership, defined as an exchange between a leader and a follower wherein the leader is satisfied with an outcome that fulfills the particular contractual obligation. Transactional leadership emphasizes the interdependence between reward and punishment. As long as a task is performed according to expectations, there is a reward; if the subordinate deviates from expectations, the response is punishment. Transactional leadership presumes little, if any, understanding or desire to know why things are done or not done in a particular manner. There is very little emphasis on the needs of the followers or their personal development.

Transformational and Transactional Leadership

- Transformational and transactional leadership can both be ethically based.
- They reflect two different value systems and motives.
- Transformational leaders possess an organic worldview characteristic of a deontological perspective. Such a perspective asserts that leaders' actions have a morally intrinsic value. An act is moral when it, as a means to an end, is executed with a sense of duty to others guided by Kantian pure reason; this is referred to as genuine or moral altruism.
- Transactional leaders reflect teleological ethics in which ends or outcomes are emphasized. The leader has an atomistic worldview, meaning that he/she feels separate from others and emphasizes independence rather than interdependence.

(Kanungo, 2001)

CAREER HIGHLIGHTS

Transformational Approach

Chief Reuben Greenberg
Charleston Police Department, South Carolina

This phenomenon of personal vision and transformation appears to be the case with Chief Reuben Greenberg of the Charleston Police Department. He introduced his values and beliefs to the Charleston Police Department and that, combined with his long and unprecedented tenure in this organization, enabled him not only to introduce change but also to make it happen and maintain it for long enough that the organization became identifiable with him and his personality rather than with the function that it serves in the community.

Since 1982, Greenberg has been chief of the Charleston Police Department, a force with approximately 350 officers ("Greenberg hinting," 1999).[1] He officially retired in 2002 but has kept working as chief under a program that allows state employees to work past their retirement and earn both their salary and retirement benefits ("Like he never left," 2002). By most accounts, Greenberg is a popular chief with national prominence (Bauer, 1999; Munday, 2002; Stevens, 2001). A Charleston synagogue even dedicated a stained-glass window to the chief, likening his qualities of strength and protection to the biblical Reuben (Munday, 2002). In a snapshot of his policing philosophy in *Fast Company* magazine, Greenberg stated that he runs a service organization, the success of which he measures by complaints and compliments: "Ultimately, we don't just want to fight crime—we want to make life better for people. So all of our officers can make on-the-spot decisions to help citizens" (McCauley, 1999, p. 97).

Greenberg received his bachelor's degree from San Francisco State University in 1967 and two master's degrees from the University of California at Berkeley: public administration (1969) and city planning (1975). He has taught as an assistant professor at California State University and has been an instructor at University of North Carolina[2] and Florida International University ("Reuben M. Greenberg," n.d.).

He has held the positions of undersheriff of the San Francisco County Sheriff's Department; chief of police in Opa Locka, Florida; chief deputy sheriff of Orange County, Florida; director of public safety in Mobile, Alabama; and deputy director of the Florida Department of Law Enforcement ("Chief Reuben M. Greenberg," n.d.). He was considered for the chief of police in Atlanta (Blackmon, 1994) and the NYPD Police Commissioner (Stevens, 2001). Greenberg belongs to the National Association of Black Law Enforcement Executives and the International Association of Chiefs of Police. He has received several national awards and is the author of a 1989 book titled *Let's Take Back Our Streets* ("Chief Reuben M. Greenberg," n.d.).

Greenberg's accomplishments during his 22-year career in Charleston are numerous. He ended the city's open-air drug markets by effectively placing uniformed police officers on street corners where the drugs were sold ("Adding police," 1997). However, he has expressed frustration at the courts' response to drug crimes, especially in high-crime areas such as the East Side neighborhood of Charleston. Greenberg implemented foot and horseback patrols in the area, along with drug-sniffing dogs, which have been successful in making 900 arrests a year for narcotics possession or resale (Hicks, 2000). He also introduced advanced technology to better respond to firearms offenses, in particular, acoustic sensor technology, which uses computers and phone lines to pinpoint where guns have been fired in the city (Smith, 2003). Yet he has realistically pointed out that such community-oriented efforts probably did not eliminate such crimes but merely displaced them to other jurisdictions. "Displacing crime means displacing it from here to somewhere up north. How far north it goes is no concern of ours," he was reported as saying in 1999 (Menchaca, 1999, p. A1).

Hiring practices also reportedly improved under Greenberg. Five police officers were unable to read or write when Greenberg became chief of police. Since then, he has raised recruitment standards, accepting only college graduates for officer positions. He has also emphasized military discipline in training new officers by requiring recruits to have excellent physical fitness (Menchaca, 1999). In addition, he has also maintained a ratio of black officers that reflects Charleston's population: 50% of the officers on the force are black; in Charleston, 41% of the citizens are black (Jacobs, 1999).

Initiatives related to juveniles have also characterized Greenberg's approach. In 2001, he established an informal curfew to protect youths during a summer spike in crime. Police were instructed to escort home anyone under age 17 found outside after 11 p.m. (Hardin, 2001; Lundy, 2002). A truancy policy has officers pick up juveniles and take them home before they commit a crime (Bauer, 1999). Moreover, the Charleston Police Department has teamed up with the Charleston Yacht Club to provide sailing memberships and instruction to disadvantaged youths. The program was designed to promote an interest in sailing and provide an alternative to delinquent activities (McCormack, 2002).

Among his controversies, Greenberg has been criticized for his approach to issues of importance to the black community. A police officer shot a black schizophrenic man during a confused struggle in a dark hallway of the victim's house on May 21, 1999; a white officer accidentally shot and killed a black officer who had also responded to the call. The Southern Christian Leadership Conference called their deaths an example of police racism and held Greenberg accountable. The officer involved, however, was cleared of any wrongdoing in a subsequent police investigation (Fennell, 1999). Charleston has no civilian review board (Jacobs, 1999).

Greenberg's positions on racial profiling are also controversial. Greenberg told the press that most police stops of blacks are reasonable because blacks commit an inordinate amount of violent crime, although he stated that blacks are also the most often victimized. According to Greenberg, the problem is how innocent blacks are treated when they are stopped. "The problem is we don't know how to disengage. Most officers, once convinced that they have the wrong guy, just tell the citizen, 'Hey, beat it; get out of here'" (Jacobs, 1999, p. A1). This response failed to improve his public relations because critics pointed out that in 80% of stops for a felony offense, the stopped person turns out not to be the suspect (Jacobs, 1999). Greenberg continued to minimize the importance of racial profiling when the department was accused of the practice in subsequent forums, including forums hosted by black and progressive groups (Brazil, 2001), and at a Cato Institute debate where he stated, "So-called racial profiling . . . does not have any appreciable impact on our criminal justice system" (Piacente, 2001, p. A1). He also accused opponents of racial profiling of being more interested in weakening drug laws and not augmenting the rights of blacks. The comment angered both liberals and conservatives. Greenberg had hoped to reframe the debate in terms of the problem he believes is more relevant: black-on-black crimes (Piacente, 2001). In addition, Greenberg opposed federal hate crimes legislation and told a congressional panel that the existing laws adequately provide for arrest and prosecution of offenders involved in bias-related incidents (Piacente, 1999).

In other conflicts, a protracted dispute between the police department and residents of James Island led to an accusation of harassment against Greenberg personally. A regular at the Riverland Terrace Boat Landing filed an incident report against Greenberg for allegedly harassing him for drinking beer at the park. Greenberg contended that he was not harassing people at the boat landing but was enforcing city codes prohibiting alcohol consumption in public places (Rucker, 1995).

Another controversy ensued when Greenberg made a comment, reported in the press, that the family of a 71-year-old Alzheimer's sufferer who wandered from his home was not properly caring for him. "You ought to get a warrant and have those SOBs arrested," he was reported as saying. The following day Greenberg declined State Senator Robert Ford's request for an apology on behalf of the family ("Greenberg's remarks," 1995).

Greenberg has also had differences with Charleston's county sheriff. An impasse between the two law enforcement agencies began in December 1992 when the two were observed in an angry dispute over a hazardous chemical leak. In addition, they have had some jurisdictional arguments stemming from the city's attempt to annex unincorporated parts of the county, thereby shrinking the territory patrolled by the sheriff's officers ("Charleston chief," 1995). However, the two reportedly agree on Greenberg's stance on racial profiling (Porter, 2002).

Greenberg publicly endorsed Elizabeth Dole during her 2000 presidential election campaign. He indicated that her gun-control position was appealing because it neither infringed on 2nd Amendment rights nor ignored the need to protect cops and children. Dole's plan included mandatory trigger locks on guns, background checks for gun-show purchases, and bans on assault rifles (Kropf, 1999). In addition, Greenberg started a program offering $100 to anyone who turns in someone with an illegal firearm (Bauer, 1999). However, the endorsement of Dole's gun-control policies seemed to be contradicted later when Greenberg implied to Charleston business owners that they could best protect themselves by procuring firearms, calling burglaries and robberies "a fact of life" (Walker, 2003, p. B1).

Transformational leadership can be truly implemented only through a long-lasting tenure and influence on all aspects of an organization. This influence, as in the case of Greenberg, frequently extends from selection and recruitment through training and changes both in the organizational structure and in the subculture of the employees. Within the rigid environment of police organizations, such profound changes can be achieved only through a long and uninterrupted "reign." On this author's first visit to research the Charleston Police Department, the taxi driver from the airport commented on my destination of the police department. His reaction was indeed very telling: "Yes, I know Reuben. Please say hello from me—Reuben *is* the Charleston Police Department." No other statement can characterize better a truly transformational police leader.

Reuben Greenburg's tenure as Chief of the Charleston Police Department came to a sudden end due to the events and fallout of an incident on August 7, 2005 (Parks, 2005). On August 7, 2005, a woman called the police on Greenberg who was driving his police cruiser erratically and appeared to be drunk (Parks, 2008). Hearing the dispatch on the radio, and realizing that he was the one being referred to, Greenberg pulled the woman over and confronted her (Parks, 2008). He was slurring his speech during the encounter (Parks, 2008). Another responding officer, a County Sheriff arrived on the scene and the woman told the Sheriff that Greenberg banged on her car with his fists (Parks, 2008). Greenberg denied the accusation; however, the County Sheriff reported seeing knuckle prints on the woman's car door (Parks, 2008).

Over the next nine days, it was discovered that Greenberg had had three prior strokes and that his slurred speech was the result of those strokes and not from being drunk (Parks, 2008). It was also discovered that Greenberg's doctor had been pressuring him to retire for at least six months prior to the incident (Parks, 2008).

On August 16, 2005, he retired from the Charleston Police Department for medical reasons (Parks, 2008). After retiring, Greenburg moved to Tryon, North Carolina where he delivered meals for the Meals on Wheels program and taught handicapped children to ride horses (Parks, 2008). He also worked as part of the security team in his towns annual barbecue festival (Parks, 2008). Greenberg is now deceased.

---

[1] Greenberg was still chief as of August 2004.
[2] While teaching at UNC, Greenberg was a mentor figure to Charles A. Moose, encouraging him to consider a career in policing ("Greenberg mentored," 2002).

CAREER HIGHLIGHTS

Transactional Approach

Chief Daryl Gates
Los Angeles Police Department, California

Daryl Gates was born on August 30, 1926, and was raised in Glendale, California. In 1945, he served in the U.S. Navy in World War II. Upon returning home after the war, he attended college at the University of Southern California on the GI Bill but did not finish (Gates & Shah, 1992). He served as the chief of police of the LAPD for 14 years, from 1978 to 1992.

Gates joined the LAPD in September 1949. After going to the academy, he worked in accident investigation and then was transferred to patrol (Gates & Shah, 1992). Gates was mentored by then-LAPD Police Chief William H. Parker. In 1950, he became Parker's chauffeur, and their relationship blazed a trail for Gates's career in policing (Shapiro, 1992). In 1965, as a patrol commander when the Watts riot erupted, Gates personally led lines of officers through the upheaval in order to resume control. After the riot, he was critical of the department for not providing officers with the tools and leadership to deal with civil unrest (Shapiro, 1992).

Gates became chief of the LAPD in 1978 at a time when the department's budget was particularly slim. By 1985, the department had only 6,900 sworn personnel, a ratio of 1 for every 429 inhabitants (New York and Chicago, at the time, had ratios of approximately 1 officer for every 242 inhabitants).[1] At the same time, the population of the city was exploding due to the immigration of hundreds of thousands of Hispanics and Asians to Los Angeles ("History," n.d.).

Among Gates's accomplishments was the master plan he developed for security at the 1984 Olympic Games held in Los Angeles ("Former Los Angeles police chief," 2003), which elevated the LAPD to being perceived as one of the top police forces in the country (Stone, 1991).

Moreover, Gates created the first Special Weapons and Tactical Team (SWAT), now a common feature of most urban police departments (Stone, 1991). Whereas in the context of SWAT,

Gates's aggressive police style was praised, this style was also viewed suspiciously in the context of the battle against gangs in the city. For example, in 1988, a drug raid in south Los Angeles involving more than 80 officers turned into a rampage on four apartments where residents were harassed and beaten. The raid turned up only one ounce of cocaine and six ounces of marijuana. Subsequent lawsuits led to city settlements of $3.5 million (distributed among 52 litigants). A report on the incident, released to the public in 1991, revealed that the gang unit operated without discipline. Officers assigned to the gang task force were uncertain about who was in charge, and commanders were unsure about how to operationalize their authority. None of the operations devised were routinely reviewed by senior officers, nor were the supervisors subject to performance evaluations. Furthermore, the report documented that the war on gangs was set up in a guerrilla-like military fashion and promoted a warlike mentality among officers ("Report," 1991).

Gates founded the Drug Abuse Resistance Education (DARE) program in cooperation with the Los Angeles public schools in 1983 ("Former Los Angeles police chief," 2003). The program involved 17 one-hour sessions with a police officer who provided information about illegal drug use and its consequences as well as strategies to resist drugs. The program eventually was adopted in 70% of schools across the United States and has spread to 44 countries. However, empirical evaluations of the program suggest that there has been no significant reduction in actual drug use as a result of the program (Walker & Katz, 2002; Wyson et al., 1994).

Gates is credited with initiating community policing in Los Angeles. However, Goldstein (1990) indicates that one initiative in 1985 in the Wilshire area, the Community Mobilization Project (CMP), eventually reverted to traditional policing. Pressure to respond to calls for service undercuts the problem-oriented approach (Goldstein, 1990), perhaps because the LAPD was stretched thin. However, media sources in the early 1990s reported that Gates's community-policing programs appeared to be effective in combating crime. As a result, during this time, Gates was a favorite of Washington politicians and was often mentioned by then-President George H. W. Bush as an effective crime fighter ("Beating crime," 1991).

Gates was known throughout his career as being prone to making inflammatory, discriminatory, and racist remarks. Within the first few weeks of his tenure, Gates stated that Hispanic officers were not being promoted at the same rate as others because they were "lazy" (Stone, 1991). In 1982, he justified the use of now-banned police choke holds that had led to the deaths of several black suspects, saying, "We may be finding that, in some blacks, when [the choke hold] is applied, the veins or arteries do not open up as fast as they do in normal people" (Domanick, 1991, p. 39). He told reporters in 1988 that a victim of a police beating was "probably lucky that's all he got." In 1990, he stated that casual drug users should be shot. The following year he called an immigrant cop killer "an El Salvadoran drunk—a drunk who doesn't belong here" (Stone, 1991, p. A9).

The most notorious incident during Gates's tenure as chief was the March 3, 1991, police beating of black motorist Rodney King, which ultimately ignited the April 29, 1992, Los Angeles riots. King, who was on parole after being convicted of robbery, was beaten 56 times with clubs and stunned with a 50,000-volt dart gun while being arrested for drunk driving. The beating was caught on videotape that showed four officers attacking King while another 21 watched, despite a crowd of citizen onlookers ("Beating crime," 1991). Gates's initial response to the incident was to order a review of all recent police brutality complaints and to have department psychologists reinterview officers involved in such cases ("Probes," 1991, A3). However, he refused to accept any responsibility in the King incident and insisted that it was an "aberration" despite the fact that the city was then paying at least $10 million in claims to blacks and Hispanics who had been brutalized by LAPD officers (Kramer, 1991).

Pressure mounted for Gates to resign ("Probes," 1991, A3). Mayor Tom Bradley, a former police officer, initially was cautious about placing blame on Gates, but within weeks of the incident, he called for Gates to step down (Cannon, 1991). Gates was then ordered by the LA Police Commission (which includes the city council and the mayor) to take a 60-day paid leave of absence while an investigation was conducted into his potential mismanagement of the department. Gates complied but also said he would sue the commission ("The chief," 1991). Thus ensued a political battle to force Gates to resign. Legally, Gates could not be directly fired by the police commission absent proof of corruption or criminal behavior.[2]

The beating of King had a dramatic effect on public opinion in Los Angeles and around the country. Prior to the episode, 70% of whites in Los Angeles approved of the LAPD, but between March 1991 and May 1992, this approval rating dropped to 40%. The approval of blacks fell to only 14% in late March 1991 (Walker & Katz, 2002). According to the American Civil Liberties Union (ACLU), complaints of police brutality in Los Angeles rose from 75 per week before King's beating to 350 weekly (Cannon, 1991). Celebrities residing in the city joined with black groups in denouncing Gates specifically. For example, actor Kim Basinger proclaimed that Gates had lost the moral authority necessary to continue as police chief. However, it appears that a majority of cops within the department supported Gates ("Beating crime," 1991). It also had a national effect; the congressional black caucus and then-Attorney General Dick Thornburg initiated two federal inquiries into police brutality on March 14, 1991 ("Beating crime," 1991).

On July 9, 1991, the Christopher Commission released a 228-page report criticizing Gates and the LAPD for a pattern of brutality, including the King beating (Orlov, 1991). Officers were routinely not disciplined for excessive use of force; 63 cops had 20 or more civilian complaints. An officer who dragged a handcuffed suspect by his feet down a hallway was recommended for a position of training recruits (Turque & Foote, 1991). The commission also discovered transcripts of racist messages between patrol-car computers (Orlov, 1991), the sheer volume of which indicated that such talk was commonplace on the job. Blacks were derided as "monkeys" and gays as "bun boys." One officer described a police helicopter "attack" of Watts in which officers were reenacting the film *Apocalypse Now* and playing "Ride of the Valkyries" from patrol-car loud speakers (Turque & Foote, 1991). There was also extensive testimony from top brass that was critical of Gates's leadership (Orlov, 1991). In an indictment of Gates, Assistant Chief David Dotson told the commission that Gates "failed miserably" in holding supervisors accountable for officers' conduct. The report recommended the replacement of Gates for—at the very least—tolerating a department saturated in racism and brutality (Turque & Foote, 1991). A subsequent Amnesty International report indicated that police brutality by the LAPD could be classified as torture as it commonly rose to the level of shootings, beatings, and electric stun-gun shocks. The report conveyed that such conduct was evident even against suspects who posed little or no threat to officer safety ("Criticism mounts," 1992).

Despite calls for his removal by citizens, celebrities, interest groups, the Christopher Commission, the mayor, and the city council, Gates soldiered on. He only reluctantly agreed to retire at the end of June 1992, not before he was the focus of even more outrage following riots on April 29, 1992. Despite assuring the city that his force was ready to deal with any unrest stemming from the King verdict, the LAPD failed to control the riots, which occurred after the officers who beat King were found not guilty. After four hours of unrest by mobs of looters and arsonists, the police had lost control of South Central Los Angeles; motorists trapped in one intersection were dragged from their cars and assaulted (Shapiro & Curry, 1992). Four days of violence led to 55 deaths and 2,000 injuries ("LAPD has changed," 2002). At the time the riots broke out, Gates was attending a political fund-raiser in Brentwood to raise money to challenge a ballot initiative that would make the police chief more accountable to elected officials ("L.A. officials," 1992; Shapiro, 1992; Shapiro & Curry,

1992). Dotson described to reporters Gates's lack of leadership in preparation for the verdict and the potential unrest. Gates reportedly refused to clarify issues around the chain of command and claimed to have a plan for dealing with unrest, yet none was shared with his commanding officers. During the first 90 minutes of the riot, Gates indicated he was in touch with his commanding officers, but this was contradicted by other accounts. In addition, there was no communication in the first hour with fire department crews waiting for police escorts (Shapiro & Curry, 1992). A report released by former FBI and CIA Director William Webster in November 1992 concurred that there was no real evidence of a workable plan to respond to the crisis. Most of the criticism in the report was leveled at Gates for misleading city officials to believe the city was ready for the verdict. Reportedly Gates, even before reading the report, called Webster and his team "liars" ("Caught off guard," 1992).

After the riots, the police commission pursued disciplinary action against Gates in order to force him out ("L.A. officials," 1992). They also secured a new police chief to follow Gates: Willie Williams, then police commissioner in Philadelphia (Tharp & Friedman, 1993; Yang, 1992). Although Gates had agreed to the June resignation, early in that month he threatened to stay on in order to spite the commission ("L.A. officials," 1992). That move would have jeopardized the appointment of Williams (a delay, due to the city charter, would effectively nullify Williams's appointment and force a new chief selection process; Yang, 1992). Gates told the *Los Angeles Times*, "I said I was going to retire at the end of June and my feeling is now, 'Screw you. I'll retire when I want to retire'" ("L.A. officials," 1992, p. A3). A couple of days later, he denied that he was serious about the threat, stating that although he wanted to hurt the city officials, he had given his word to Williams that he was retiring (Yang, 1992).

In August 1992, Gates's autobiography, *Chief: My Life in the LAPD*, was released. It was written after the King beating but before the riots (Shapiro, 1992). Since his departure from the LAPD, Gates has maintained a public voice in criminal justice issues. At a 1996 pro-police rally in Los Angeles (following a police beating of undocumented immigrants who led police on an 80-mile chase), he spoke about the need for aggressive policing (Babwin, 1996). Gates continues to deny wrongdoing in the King beating, telling *ABC News* in 2001 that King "got whacked a few extra times" but "brought it on himself" (Werner, 2001). In a 2002 *Frontline* interview, he did concede that the King beating "looked horrible, no question about it" but claimed that it was not a racially motivated incident ("Interviews," 2001). Gates continues to defend the LAPD as the best police department in the world, but one thwarted by vindictive city politicians (Werner, 2001).

Since 1992, Gates has worked as a law enforcement consultant (Werner, 2001). In March 2003, he was named senior advisor of SAFKEY Corp., which produces a security system to prevent car theft, carjacking, and drunk driving ("Former Los Angeles police chief," 2003).

Gates, a driven leader, had a vision of what was right and wrong for the LAPD; however, this vision was implemented through assignments of tasks to individuals without particular attention given to the match between the people and these tasks. As long as the tasks were performed and accomplished, the officers were rewarded. When the tasks were not performed according to the vision, the officers were punished, but this exchange was based more on the classical transactional approach of cost-benefit exchange process than on the analysis of the real price of this exchange. A transactional leadership style can lead to ethically questionable behavior on the part of a leader. Gates's lack of sensitivity toward minorities exhibited itself in the fact that officers accused of inflammatory and racist remarks remained on the force instead of being swiftly reprimanded, disciplined, and (if necessary) removed. His lack of sensitivity toward minority communities was one of the biggest mistakes of his career, particularly when he underestimated the reaction of minority communities in the aftermath of the Rodney King incident.

But for Gates it was about a transactional exchange. Transactional leaders frequently engage in a laissez-faire type of behavior, not interfering with subordinates and empowering them in the decision-making process—as long as they deliver the desired

results. His overreliance on the transactional approach led to too much empowerment of the commanding officers during the Rodney King incident and to the eruption of the riots. Leaders who use the transactional style of leadership do not become more effective regardless of whether they have the capacity to reason morally. This is because, by definition, transactional leadership is merely a dyadic exchange to satisfy individual needs (Turner et al., 2002). Gates's individual needs were closely tied to his view of what the right thing was for the LAPD; however, he never took a larger or more collaborative view of what it was that the force really needed and ultimately demanded the transaction without a more in-depth understanding of the outcomes.

After leaving the LAPD, Gates worked as a radio talk show host and as a consultant for several companies (Daryl, 2010). He also helped create the SWAT video game series (Winton & Blankstein, 2010). Gates also played the Los Angeles Police Chief in the 2008 film Street Kings (Winton & Blankstein, 2010). Gates died of bladder cancer on April, 16, 2010 (Daryl Gates, 2010; Orlov, 2010; Winton & Blankstein, 2010).

---

[1] New York had a force of 28,700 and a population of approximately 7 million; Chicago had a force of 12,399 and a population of approximately 3 million.
[2] A 1937 amendment to the city charter provided for a virtual lifetime appointment of the police chief, who could not be fired barring serious malfeasance. This policy was created as a way of insulating the police chief from the whims of politicians. At that time, police officers were dependent on political patronage (Witkin & Tharpe, 1992).

Exercise

The police chiefs detailed in this chapter have been examined through the prism of transformational theory. Using what you have learned from other theories in this book or from the summaries (as indicated in the Introduction) you have read thus far, it is possible to consider their approaches through the lens of different theories. What other theories can you find applicable to these chiefs? Are there lessons to be learned from those applications, or do those theories indicate alternative paths or approaches that might have yielded a more successful outcome?

References

Adding police is a priority (1997, January 28). *The Atlanta Constitution*, p. 10A.

Babwin, D. (1996, April 16). Deputies defended by Gates, crowd: The former LA police chief blasts media's handling of the beatings of undocumented immigrants. *The Press Enterprise*, p. A1.

Bass, B.M. (1985). *Leadership and performance beyond expectations*. New York: Free Press.

_____(1999). Two decades of research and development in transformational leadership. *European Journal of Work and Organizational Psychology*, 8 (1), 9–32.

_____& B.J. Avolio (1994). *Improving organizational effectiveness through transformational leadership*. Thousand Oaks, CA: Sage.

Bauer, T. (1999, February 23). Forum tackles policies. *The Augusta Chronicle*, p. C8.

Beating crime (1991, March 23). *Economist*, 318 (7699), p. 28.

Bennis, W.G. & B. Nanus (1985). *Leaders: The strategies for taking charge*. New York: Harper & Row.

Blackmon, D.A. (1994, November 17). Harvard's rivals? None came close. *The Atlanta Constitution*, p. C2.

Brazil, B. (2001, February 27). Minority residents accuse officers of racial profiling. *The Post and Courier*, p. B3.

Bryman, A. (1992). *Charisma and leadership in organizations*. London: Sage.

Burns, J.M. (1978). *Leadership*. New York: Harper & Row.

Cannon, L. (1991, April 8). Anti-police atmosphere as thick as smog over L.A. *Seattle Post-Intelligencer*, p. A9.

Caught off guard: A report on the L.A. riots singles out the city's former top cop as culprit (1992, November 2). *Time*, 18, p. 140.

Charleston chief, sheriff feuding over just about everything (1995, October 1). *The Herald*, p. 4B.

Chief Reuben M. Greenberg (n.d.). Retrieved May 29, 2003, from www.lsrcl.org/Bios/greenburgbio.htm.

Conger, J. (1999). Charismatic and transformation leadership in organization: An insider's perspective on these developing streams of research. *Leadership Quarterly*, 10 (2), 145–179.

_____& R. Kanungo (1998). *Charismatic leadership in organizations*. Thousand Oaks, CA: Sage.

Criticism mounts as Gates leaves police force: King arrested (1992, June 27). *Seattle Post-Intelligencer*, p. A3.

Daryl Gates, L.A's police chief during '92 riots, dies. (2010, April 16). *USA Today*. Retrieved May 31, 2011, from http://www.usatoday.com/news/nation/2010-04-16-daryl-gates-dies_N.htm.

Daryl F. Gates, 83; police chief during Rodney King riots. (2010, April, 17). *The Washington Post*. Retrieved May 31, 2011, from http://www.washingtonpost.com/wp-dyn/content/article/2010/04/16/AR2010041604655.html.

De Vries, R.E., R.A. Roe & T.C.B. Taillieu (1999). On charisma and the need for leadership. *European Journal of Work and Organizational Psychology*, 8 (1), 109–133.

Domanick, J. (1991, July/August). Field marshal Daryl Gates. *Mother Jones*, 16 (4), 39.

Dvir, T. & B. Shamir (2003). Follower development characteristics as predicting transformational leadership: A longitudinal field study. *Leadership Quarterly*, 14 (3), 327–344.

_____, E. Dov, B.J. Avolio & B. Shamir (2002). Impact of transformational leadership on follower development and performance: A field experiment. *Academy of Management Journal*, 45 (4), 735–744.

Fennell, E.C. (1999, June 29). Officer won't face charges. *The Post and Courier*, p. A1.

Former Los Angeles police chief Daryl Gates has a new gig (2003, March 11). *City News Service*.

Fuller, J.B., C.E. Patterson, K. Hester & D.Y. Stringer (1996). A qualitative review of research on charismatic leadership. *Psychological Reports*, 78, 271–287.

Gates, D. & D.K. Shah (1992). *Chief: My life with the LAPD*. New York: Bantam Books.

Giampetro-Meyer, A., S.J.T. Brown, M.N. Browne & N. Kubasek (1998). Do we really want more leaders in business? *Journal of Business Ethics*, 17, 1727–1736.

Goldstein, A. (1990). *Problem-oriented policing*. New York: McGraw-Hill.

Greenberg hinting at departure (1999, November 17). *The Post and Courier*, p. A1.

Greenberg mentored police chief (2002, October 11). *The Post and Courier*, p. A9.

Greenberg's remark sparks angry letter (1995, December 10). *The Post and Courier*, p. A27.

Hardin, J. (2001, July 17). Cops, city want teens off streets. *The Post and Courier*, p. A1.

Hicks, B. (2000, April 12). East Side seeks more police. *The Post and Courier*, p. B7.

History of the LAPD (n.d.). Retrieved July 7, 2003, from www.lapdonline.org/general_information/history_of_the_lapd/gen_history_lapd.

House, R.J. (1976). A 1976 theory of charismatic leadership. In J.G. Hunt & L.L. Larson (Eds.), *Leadership: The cutting edge* (pp. 189–207). Carbondale, IL: Southern Illinois University Press.

_____ & B. Shamir (1993). Toward the integration of transformational, charismatic, and visionary theories of leadership. In M. Chemers and R. Ayman

(Eds.), *Leadership: Perspectives and research directions* (pp. 81–107). New York: Academic Press.

Interviews: Daryl Gates (2001, February 27). *Frontline*. Retrieved on July 1, 2003, from www.pbs.org/wgbh/pages/frontline/shows/lapd/interviews/gates.html.

Jacobs, D. (1999, September 14). S.C. police chief: Racial profiling not the problem. *Knoxville News-Sentinel*, p. A1.

Judge, T.A. & J.E. Bono (2000). Five-factor model of personality and transformational leadership. *Journal of Applied Psychology*, 85 (5), 751–765.

Kanungo, R.N. (2001). Ethical values of transactional and transformational leaders. *Canadian Journal of Administrative Sciences*, 18 (4), 257–265.

Kramer, M. (1991, April 1). Gates: The buck doesn't stop here. *Time*, 137 (13), 25.

Kropf, S. (1999, July 29). Greenberg endorses Mrs. Dole. *The Post and Courier*, p. B3.

Kuhnert, K.W. & P. Lewis (1987). Transactional and transformational leadership: A constructive/developmental analysis. *Academy of Management Review*, 12, 648–657.

L.A. officials want police chief dismissed (1992, June 8). *Seattle Post-Intelligencer*, p. A3.

LAPD has changed in decade since rioting (2002, April 29). *USAToday.com*. Retrieved July 7, 2003, from www.usatoday.com/news/nation/2002/04/28/riots.htm.

Like he never left (2002, July/August). *Law Enforcement News*, vol. XXVIII, nos. 581, 582. Retrieved August 10, 2004, from http://www.lib.jjay.cuny.edu/len/2002/ 07.08/p&p.html.

Lowe, K.B., K.G. Kroeck & N. Sivasubramaniam (1996). Effectiveness correlates of transformation and transaction leadership: A meta-analytic review of the MLQ literature. *Leadership Quarterly*, 7, 385–425.

Lundy, S. (2002, March 17). More of a police presence appreciated by residents. *The Post and Courier*, p. 10A.

Masi, R.J. (2000). Effects of transformational leadership on subordinate motivation, empowering norms, and organizational productivity. *International Journal of Organizational Analysis*, 8 (1), 16–47.

McCauley, L. (1999, May). Measure what matters. *Fast Company*, 24, p. 97.

McCormack, F. (2002, June 6). Program teaches kids adventures of sailing. *The Post and Courier*, p. B1.

Menchaca, R. (1999, February 25). Greenberg praises department. *The Post and Courier*, p. A1.

Munday, D. (2002, June 28). Synagogue window honors chief. *The Post and Courier*, p. B3.

Orlov, R. (1991, July 21). L.A. chief Gates reportedly agrees to quit under fire. *The Seattle Times*, p. A1.

Orlov, R. (2010, April 16). Former LAPD Chief Daryl Gates dies. *Los Angeles Daily News*. Retrieved May 31, 2011, from http://www.dailynews.com/news/ci_14901214.

Parks, N. (2008, April 13). Where's Reuben? Former Charleston police chief opens up about his retirement in North Carolina mountains, health issues that led to his departure. *The Post and Courier*. Retrieved May 31, 2011, from http://www.postandcourier.com/news/2008/apr/13/wheres_reuben_former_charleston_police_ chief_opens/.

Piacente, S. (1999, August 5). Greenberg says hate crimes laws are not needed. *The Post and Courier*, p. B1.

_____ (2001, May 20). Racial profiling remarks set off Washington buzz. *The Post and Courier*, p. A1.

Porter, A. (2002, April 21). Charleston county sheriff defends use of racial profiling. *The Post and Courier*, p. A1.

Probes of police brutality: In L.A., pressure grows on chief to resign (1991, March 21). *Seattle Post-Intelligencer*, p. A3.

Report of command breakdown in police war on LA gangs (1991, June 24). *The San Francisco Chronicle*, p. A15.

Reuben M. Greenberg (n.d.). Retrieved May 29, 2003, from www.charlestonpd.org/rmgbio.html.

Rucker, A. (1995, August 9). Greenberg accused of harassment. *The Post and Courier*, p. A1.

Shamir, B., E. Zakay, E. Breinin & M. Popper (1998). Correlates of charismatic leadership behavior in military units: Subordinates' attitudes, unit characteristics, and superiors' appraisal of leader performance. *Academy of Management Journal*, 41, 387–409.

Shapiro, J.S. (1992, August 20). The law and Daryl Gates: No hail to this chief. *The Recorder*, p. 9.

Shapiro, W. & T. Curry (1992, May 18). Lessons of Los Angeles. *Time*, 139 (20), 38.

Smith, B. (2003, March 14). New gunfire detection system to begin in Charleston. *The Associated Press*, state and local wire.

Stevens, K. (2001, October 8). Greenberg considered for top job at NYPD. *The Post and Courier*, p. B3.

Stone, K. (1991, July 10). Conflict has marked Gates' tenure. *Seattle Post-Intelligencer*, p. A9.

Tharp, M. & D. Friedman (1993, August 2). New cop on the block. *U.S. News & World Report*, 115 (5), 22.

The chief takes a walk—For now (1991, April 14). *U.S. News & World Report*, 110 (14).

Tichy, N.M. & M.A. DeVanna (1990). *The transformational leader* (2nd ed.). New York: John Wiley.

Turner, N., J. Barling, O. Epitropaki & V. Milner (2002). Transformational leadership and moral reasoning. *Journal of Applied Psychology*, 87 (2), 304–311.

Turque, B. & D. Foote (1991, July 22). Damned—but defiant: A scorching report of a violent and racist LAPD. *Newsweek*, 118 (4), 22.

Walker, S. & C.M. Katz (2002). *The police in America* (4th ed.). New York: McGraw-Hill.

Walker, T. (2003, January 24). Chief suggests arms can keep businesses safe. *The Post and Courier*, p. B1.

Weber, M. (1947). The theory of social and economic organizations (T. Parsons, trans.). New York: Free Press.

Werner, E. (2001, March 3). Ex-chief still defiant: A decade later, Gates said Rodney King "brought it on himself." *ABCnews.com*. Retrieved July 1, 2003, from abcnews.go.com/sections/us/DailyNews/Rodney_king_gates010303.html.

Winton, R., & Blankstein, A. (2010, February 17). Former LAPD Chief Daryl Gates is seriously ill. *The Los Angeles Times*. Retrieved May 31, 2011, from http://articles.latimes.com/2010/feb/17/local/la-me-gates17-2010feb17.

Witkin, G. & M. Tharpe (1992, June 8). Police chiefs at war. *U.S. News & World Report*, 112 (22), 33–34.

Wyson, E., R. Aniskiewicz & D. Wring (1994). Truth and DARE: Tracking drug education to graduation as symbolic politics. *Social Problems*, 41, 448–471.

Yammarino, F.J., W.D. Spangler & B.M. Bass (1993). Transformational leadership and performance: A longitudinal investigation. *Leadership Quarterly*, 41, 81–102.

Yang, J.E. (1992, June 9). Chief Gates says his threat was a bluff. *The Houston Chronicle*, p. A2.

6

Parameters for Empowerment and Trust
Style Theory

OVERVIEW OF THE THEORY

Style theory focuses on what leader's do. Here, leader behavior, plus intervening variables, leads to end result variables (Yukl, 1994). In this chapter, the example of Chief Earl Sanders of the San Francisco Police Department (SFPD) is used to illustrate the choices one chief made in managing a scandal when SFPD was accused of covering up officer misconduct. The other case study focuses on Chief Dennis Nowicki, who has also faced numerous challenges during his career at the Charlotte-Mecklenburg Police Department, but his method of handling them differed significantly from Sanders's.

OHIO STATE UNIVERSITY LEADERSHIP STUDIES

Studies conducted by researchers at Ohio State University explored the behavior of leaders. The researchers compiled a list of approximately 1,800 types of leadership behavior, narrowing the list down to 150 leadership functions. In the study, participants identified two broad dimensions of behavior: having consideration and initiating structure (Yukl, 1994). According to Northouse (2004), having consideration refers to the degree to which the leader is concerned for the well-being of subordinates; this is often referred to as being relationship-oriented. Initiating structure addresses the extent to which the leader focuses on goal attainment; this is referred to as task-oriented behavior.

Having consideration and initiating structure were seen as independent behavior categories, meaning that leaders could fall anywhere on the spectrum of high to low in respect to having consideration, irrespective of their orientation relative to initiating structure (Misumi, 1985; Yukl, 1994).

MICHIGAN UNIVERSITY LEADERSHIP STUDIES

These studies focused on the effect of leaders on measures of group performance. A series of field studies with leaders compared the behavior of effective and ineffective leaders in terms of productivity leading to the formulation of two dimensions of leadership behavior: task-oriented behavior and relationship-oriented behavior (Yukl, 1994).

Style Theory

- What leaders do (not who they are).
- Task-oriented behavior vs. relationship-oriented behavior.
- Universal theories of leadership: Most effective leaders earn a 9,9 (the highest possible score on a grid).
- Both task-oriented and relationship-oriented: Most effective leaders change the situation rather than adapt to it.

Cartwright and Zander (1960) refer to these functions as 1) leaders' concern with the achievement of a group goal and 2) leaders' maintenance and strengthening of the group itself. Task versus relationship behavior was conceptualized along a continuum (and therefore being high on one dimension meant being low on the other), as described in Northouse (2004). A third dimension of leadership is participative leadership.

Participative Leadership

In addition, the Michigan University studies proposed a third style category called participative leadership which encompasses leaders who use group supervision rather than supervising each employee separately (Kahn, 1956; Yukl, 1994). With participative leadership, the supervisor instills in the employees a sense of responsibility as to the organizational goals and does not micromanage each small specific task. Ostensibly, this approach is better fostered when the group also has high peer loyalty. Using group meetings to solve problems and giving the employees a sense of ownership of the problems (and a hand in their solution) are also important facets of this approach (Likert, 1961).[1] The joint decision-making aspect of participative leadership can be seen as a continuum with four main categories (Yukl, 1994):

1. *Autocratic.* The leader makes decisions without soliciting any feedback from the subordinates.
2. *Consultation.* The leader solicits subordinates' opinions but makes the decision alone.
3. *Joint.* The leader discusses the decision with subordinates, and collectively the group makes the final choice, with the leader having no more influence on the choice than subordinates.
4. *Delegation.* The leader gives the group authority to make the decision, specifying only the outer limits on the choice. Prior approval in this case may not be required before implementing the decision.

Kahn (1956) places particular emphasis on delegation, noting that delegation affords the leader greater respect from subordinates. However, as Yukl (1994) notes:

Sometimes what appears to be participation is only pretense . . . a manager may solicit ideas and suggestions from others, and then ignore them when making a decision. Likewise, the manager may ask subordinates to make a decision, but do so in such a way that the subordinates are afraid to show initiative or deviate from the choice they know the boss prefers. (p. 84)

---

[1] In the Ohio State University studies, this dimension was incorporated within the notion of consideration (Yukl, 1994).

Bowers and Seashore (1966) came to a similar conclusion as the Michigan researchers that some form of participative leadership style correlates with effectiveness:

> [B]usiness growth is high when the agent force does *not* hold classical business ideology; when regional managers, by accepting the opinions and ideas of their agents, encourage professional development; and when managers reduce rivalries among agents by encouraging their interaction. (p. 261)

Further, unlike prior leadership studies, Bowers and Seashore (1966) paid attention to informal leaders, making leadership independent of position in the organization.

HARVARD UNIVERSITY STUDIES

Researchers at Harvard University have conceptualized two types of leaders: task leaders and social-emotional leaders. The types parallel the task-oriented and relationship-oriented styles postulated by the Michigan researchers. In one study by Zelditch (1955), the male adult in families often specialized in providing task leadership, while the female adult functioned as the maintenance, or social-emotional, leader (Reddin, 1967). This distinction seems to parallel the paternalism/maternalism dichotomy often discussed in relation to the managerial grid.

THE MANAGERIAL GRID

The managerial grid was designed by Blake and McCanse (1991; as cited in Northhouse, 2004) as a tool for leaders to assess their strengths and weaknessess. It maps behavior along two axes: concern for production (task orientation) and concern for people (relationship orientation)

Managerial Grid

1,1 (low task, low relationship) Impoverished management. Exertion of minimum effort to get required work done is appropriate to sustain organization management. It may indicate fear of termination (negative) or desire to please (positive).

1,9 (low task, high relationship) Country club management. Thoughtful attention to the needs of the people for satisfying relationships leads to a comfortable, friendly organizational atmosphere and work tempo. It may indicate fear of rejection (negative) or desire to please (positive).

9,1 (high task, low relationship) Authority-compliance management. Efficiency in operations results from arranging conditions of work in such a way that human elements interfere to some degree. It may indicate fear of failure (negative) or desire for control, mastery, and domination (positive).

5,5 (middle task, middle relationship) Middle-of-the-road management. Adequate organizational performance is possible through balancing the necessity to get work out with maintaining morale of people at a satisfactory level. It may indicate fear of humiliation (negative) or desire to belong (positive).

9,9 (high task, high relationship) Team management. Work accomplishment is from committed people; interdependence through a common stake in organizational purpose leads to relationships of trust and respect. It may indicate fear of selfishness (negative) or desire for fulfillment through contribution (positive).

(Blake and McCanse, 1991 [as summarized by Northouse, 2004])

(Northouse, 2004). These factors are mapped on intersecting axes, where minimum concern receives the score of 1 and maximum concern receives a score of 9. Therefore, a score of 1,9 would reflect a manager with a low concern for production and a high concern for relationship.

The managerial grid also provides a framework for understanding one's personal motivations, adding a third dimension to the task versus relationship grid, to show the positive or negative motivations that lead to that style. For example, a leader who is high in both task and relationship (9,9) may be motivated along a continuum of fear of selfishness (negative) to desire for fulfillment through contribution (positive). Furthermore, Blake and McCanse (1991) envision a grid for leaders' subordinate counterparts. The vertical and horizontal dimensions for subordinates are concern for the boss and concern for task accomplishment, respectively.

Finally, Blake and McCanse (1991) suggest that several factors are determiners of leadership style: organizational culture, the leader's personal values, the leader's personal history, and the assumption that there is no choice but to operate in a certain way.

STYLE AS A UNIVERSAL THEORY

This theory states that the same style of leadership is effective in all situations. Most deem the high-high leaders (9,9) the most effective because they are both relationship-oriented and task-oriented (Misumi, 1985; Yukl, 1994). Blake and Moulton (1982) and Dainty (1986) also argue for one style being the most effective, a style in which the leader changes the situation rather than adapting to it. This style is again one that is high in both task and relationship orientations. Leaders who integrate both fundamental orientations can effectively use the full range of requirements in any managerial position (Bass, 1990; Stogdill, 1974).

Other researchers have considered possible situational antecedents to leaders' adoption of one or both of the fundamental styles. Leaders who have immediate external pressure to produce desired outcomes are more likely to be task-oriented and less likely to nurture relationships. Similarly, Stogdill (1974) concludes that consideration and structure are significant "not because they are exhibited by the leader, but because they produce differential effects on the behavior and expectations of the followers" (p. 141).

Likewise, personal antecedents to behavior have also been posited, which blend into trait theory (see Chapter 12). Some have argued that people with different temperaments differentially gravitate toward either the task or the relationship orientation (Bass, 1990).

STYLE THEORY AND MILITARY LEADERSHIP

The Leader Behavior Description Questionnaire (LBDQ) was developed by researchers to gauge style and was given to military personnel in several studies in the 1950s and 1960s, as cited in Stogdill (1974; see Table 6-1). Christner and Hemphill (1955) (as cited in Stogdill, 1974) found that Air Force crews in which the commanding officer was high in consideration exhibited more friendliness, mutual confidence, conversation on duty, and willingness for combat. If the commanding officer was high in structure, however, crew members tended to exhibit only friendliness and confidence.

A study by Halpin (1954; as cited in Stogdill, 1974) found that crew members' satisfaction with their commanders during training was positively related to consideration and negatively related to structure. However, superiors viewed the commanders more positively if they were high in structure and low in consideration. During combat, crew members felt more effective when

TABLE 6-1	Style Theory in Military Environments	
During Training		**During Combat**
• Crew members' satisfaction with their commanders during training was positively related to consideration and negatively related to structure.		• Crew members felt more effective when their commanders scored high in both consideration and structure.
• Supervisors viewed the commanders more positively if they were high in structure and low in consideration.		• Supervisors of the commanders remained more positively oriented toward commanders high in structure.
• Military groups tend to be more cohesive when their leaders score high in both consideration and structure, but does cohesion translate to greater effectiveness in accomplishing organizational goals?		

(Halpin, 1954 [as cited in Stogdill, 1974])

their commanders scored high in both consideration and structure. Superiors of the commanders again were more positively oriented toward commanders high in structure in combat situations.

Giving the LBDQ to cadet flight members who were asked to describe 48 flight leaders, Newport (1962; as cited in Stogdill, 1974) showed that leaders who were equally high in consideration and structure differed from those who exhibited low levels of each style in the following ways. The high-high leaders desired individual freedom of expression, little resistance to social pressure, strong desire for power, and high levels of cooperativeness, and they were strong in aggressive attitudes. Fleischman (1957; as cited in Stogdill, 1974) found that emergence of a leader in a leaderless group task was negatively related to consideration and positively related to structure.

Stogdill (1974) concludes that military groups tend to be more cohesive when their leaders score high in both consideration and structure. However, he remains skeptical that either measure is highly related to the groups' effectiveness of accomplishing organizational goals.

STYLE THEORY AND POLICE LEADERSHIP

The only study that relates to policing and leadership styles is the Kirmeyer and Lin study (1987; as cited in Bass, 1990), which found that effective communication was best facilitated when subordinates felt they received social support from their supervisors, thereby suggesting that police

Style Theory and Police Leadership

- They studied relationships between police dispatchers and their supervisors.
- Effective communication was best facilitated when subordinates felt they received social support from their supervisors. Therefore, they responded better to managers whose style was high on the relationship-oriented continuum.

(Kirmeyer and Lin, 1987 [as cited in Bass, 1990])

dispatchers responded better to managers whose style was high on the relationship-oriented continuum (Bass, 1990).

STYLE THEORY AND THE CASES OF CHIEF EARL SANDERS AND CHIEF DENNIS NOWICKI

The style leadership theory posits that leadership effectiveness is dependent on the behavior of the leader. The two broad categories that inform such behaviors are the concern for the structure of the organization and the concern for people. According to Blake and Moulton (1982), the intersection of the two behavior dimensions forms a grid of styles of management. Each leader has a style that is used in all situations. In many aspects, it is indeed a sort of universal theory of leadership due to the fact that in all leadership situations one must consider the task to be accomplished and the people who are part of the implementation. Aspects of the careers of Chief's Earl Sanders and Dennis Nowicki are good examples to illustrate the style approach; however, each of them exhibits a different spectrum. Of course, it is important to remember that the purpose of this exercise is to illustrate the theories and that the career of each chief has varying facets that may not fall in line with these depictions. But to illustrate the theory, for Sanders the major consideration is relationships, or the people in his organization; for Nowicki it is tasks, or his police organization.

Although researchers contend that the high-high leaders (9,9) are the most effective leaders because they are high on both relationship orientation and task orientation, high-high (9,9) ratings are extremely unlikely (if not impossible), at least in police environments due to the inherent conflict between being relationship-oriented and meeting the demands of a critical public. It is probably possible to witness such orientation in for-profit environments, where the leaders of the organization are motivated, first and foremost, by the profit margin and where the limited level of accountability is skewed toward the interests of owners or shareholders. Such accountability does not approach the transparency and high standards demanded from police. Events of various natures and intensities are not just scrutinized and criticized internally by the Internal Affairs mechanisms; they are also frequently brought to light by members of the concerned communities and the ever-powerful media. Accountability leaves no room for cover-up, nor can a discretionary approach have even a hint of misconduct.

Traditionally, police leaders who choose a relationship orientation are considered to be "cops' cops," who gain the respect and trust of the organization, but often at the expense of perpetuating a code of silence. In the cases presented, the chief who responded to allegations of misconduct with task-oriented behavior appeared more ethical, insofar as he responded to scandals with the action demanded by public and media pressure. Neither style orientation has a greater inherent moral authority, but task-oriented policing may be more conducive to the accountability demands when considered through the prism of a police leader's tremendous contribution to a healthy democratic society.

CAREER HIGHLIGHTS

Relationship-Oriented Style

Chief Earl Sanders
San Francisco Police Department, California

Earl Sanders was born in Nagodoches, Texas, in 1938. He moved to San Francisco alone as a teenager after his mother died. In the 1950s, he lived in a rooming house and washed dishes at night, attending high school during the day (Reiterman & Glionna, 2003c). He joined the police department in 1964 after serving in the Army (Zamora & Lelchuk, 2002). He earned his bachelor's and master's degrees in public administration of justice (Reiterman & Glionna, 2003c).

In 1971, Sanders became a homicide detective and subsequently worked more than 300 cases (Reiterman & Glionna, 2003c). He and his partner, Napoleon Hendrix, were considered to be the best homicide investigators in the city's history. They investigated a number of prominent murders in the city (Zamora & Lelchuk, 2002).

Sanders is the founder of Officers for Justice, which initiated a lawsuit against the city on civil rights and discrimination grounds. In 1979, he testified in a federal discrimination trial about the racist jibes directed against him throughout his career by members of the department (Zamora & Lelchuk, 2002). His testimony in the federal case reportedly helped to weaken the old-boy network that had previously controlled the department ("Chief Sanders," 2003).

Sanders became the first African-American police chief in San Francisco history when he was appointed in 2002 (Reiterman & Glionna, 2003c). He inherited a 2,300-officer force, which was among the nation's worst at solving violent crimes. At that time, on average, 28% of violent crimes were solved ("An officer," 2002). According to a *San Francisco Chronicle* investigative report, the homicide unit solved only half of its cases, compared to an average of 61% solved in other urban departments. Police solved less than 40% of murders involving African-American victims between the years 1997 and 2001.

Murders involving Caucasian victims were solved 60% of the time (Van Derbeken, 2002). As a result of this report, Sanders made solving violent crime his top priority. He hired his retired partner from his days as a homicide detective, Napoleon Hendrix, to work on cold cases. He also attempted to transform promotions in the homicide unit by making them merit-based rather than seniority-based (Van Derbeken, 2002).

Sanders also increased the number of officers involved in community policing in hopes of reaching out to citizens who had become skeptical of overzealous policing ("An officer," 2002). He began a recruitment drive in order to raise the number of officers to the highest the budget allowed, 2,450 ("S.F.'s new chief," 2002).

Sanders's short term as police chief was overshadowed by a police scandal that became known as "fajitagate." The scandal stemmed from an incident that occurred on November 20, 2002. Three off-duty cops, Alex Fagan Jr. (the assistant chief's son), David Lee, and Matt Tonsing, allegedly assaulted two men, Jade Santoro and Adam Snyder, outside a bar in the Pacific Heights neighborhood (Van Derbeken & Sward, 2002a). Santoro was beaten and required hospital treatment for a laceration on his head and a broken nose. According to the victims, the attack was unprovoked. Reportedly, the officers demanded that the victims hand over a bag of take-out food, steak fajitas. When they wouldn't comply, the officers assaulted them (Van Derbeken & Sward, 2002b).

The case, which became known as "fajitagate," evolved into an enormous scandal that reached far beyond allegations of misconduct of three off-duty officers. The nexus of the case became police protection of the offenders and cover-up of the case's mishandling. It damaged the relationship between police and prosecutors and exacerbated race relations in the city. It eventually resulted in Sanders's estrangement from the force, although he was not formally dismissed.

Initial problems with the case involved mishandling of evidence collected in the early stages of the scandal. Allegedly, investigators failed to seize the clothes the officers wore during the assault, nor did they test for alcohol content in the officers' blood. In addition, the two assaulted men were not permitted to make an on-the-scene identification of the officers involved. Instead, the officers were allowed to drive away without being detained, interviewed, or tested for alcohol content in their blood (Van Derbeken & Sward, 2002a). They subsequently changed clothes and were permitted to drink large quantities of water before their blood-alcohol level was tested, four hours later ("Police scandal," 2003). On November 22, 2003, District Attorney Terence Hallinan had a meeting with senior police officials during which the police were unable to explain the problems with evidence collection during the early investigation (Van Derbeken & Sward, 2002a); later press reports alleged that these problems were a deliberate police cover-up (Van Derbeken, 2003a). Assistant Chief Alex Fagan Sr. withdrew from the investigation as his son was one of the officers, but only after the initial evidentiary mishandling (Van Derbeken & Sward, 2002a). Internal investigators encountered a wall of silence, finding members of the department hesitant to give statements (Van Derbeken & Sward, 2002b). Lieutenant Joe Dutto, who headed the internal investigation into the incident, was transferred to a less prestigious vice assignment by top officials; although the official reason given was that it occurred as part of a large-scale transfer of mid-level officers, Dutto called the transfer evidence of retaliation because he had failed to clear the officers of wrongdoing ("Chronicle," 2003; Reiterman & Glionna, 2003a, 2003b).

On February 27, 2003, the district attorney eventually indicted Sanders and Assistant Chief Alex Fagan (for conspiring to reassign Dutto, the investigator), as well as Deputy Chief David Robinson, Deputy Chief Gregory Suhr, Lieutenant Edmund Cota, Captain Greg Corrales, and Sergeant John Syme, for conspiracy to obstruct justice (by playing a more direct hand in the cover-up; Van Derbeken, 2003a). Less than two weeks later, the district attorney dropped the charges against Sanders and Fagan Sr. for lack of evidence, and the charges against the other accused conspirators were dismissed later that year ("Charges," 2003; Van Derbeken, 2003b).

Sanders responded to early criticism by defending his department's actions; rank-and-file officers also largely supported the chief and commanders ("Police scandal," 2003). In early public comments about the criticisms of the initial investigation, Sanders compared the unjustified accusations to the biblical experience of Jesus Christ and his persecutors (Van Derbeken & Sward, 2002b). After the indictment against him was dismissed, Sanders indicated in a statement that the charges reflected an attack against the criminal justice system itself and that it was one of the worst such attacks in California history (Glionna et al., 2003). In an interview a couple of days later, Sanders defended his department's internal investigation into the assault. He stated that he disagreed with the criticisms and that the behavior he was accused of did not constitute criminal conduct; he attributed the indictments to unspecified political motivations inside and outside the department (Reiterman & Glionna, 2003c).

In the early stages, Mayor Willie Brown, also an African American, was loyal to Sanders and the top commanders and did not demand a more thorough inquiry, nor did he ask the officers to step down ("Room," 2003). Mayor Willie Brown's response to the indictment, after an emergency meeting of the police commission, was that the charges were politically motivated ("Room," 2003; "San Francisco police chief," 2003). Eventually, the mayor's support for the police turned. Only weeks after the indictment, Mayor Brown expressed frustration at the scandal, criticizing Hallinan, Sanders, and Dutto (the officer who headed the Internal Affairs investigation). Brown expressed distress at the damaged relationship between the police and prosecutors. In addition, the mayor specifically attacked Sanders for asking the state attorney general to get involved (Glionna & Reiterman, 2003).

As public disquiet increased, the press revealed unsavory aspects of the department's history of handling misconduct under Sanders and exposed the personal records of some of the officers involved. *The San Francisco Chronicle* described the department as harboring a "permissive climate," which may have contributed to the scandal (Sward & Wallace, 2003). In Sanders's first six months at the helm, only one case had been referred to the police commission for a disciplinary hearing, compared to approximately a dozen cases under the former chief (Sward & Wallace, 2003). Assistant Chief Fagan was singled out by the press to indicate a larger problem with discipline in the department. In 1990, he had threatened California Highway Patrol officers during a roadside confrontation but was never disciplined after the incident. In addition, Fagan Jr. had 16 violent encounters with suspects on his record over a 13-month period ("Police scandal," 2003).

According to the media, the African-American community was split about Sanders and the scandal. To the African-American attorneys, civic leaders, and clergy of Bay Area Police Watch, the "fajitagate" case was another example of police brutality disproportionately affecting minorities. However, the NAACP reportedly defended the police in the case and did not credit claims of a cover-up. At least one African-American clergyman felt that the indictments represented an attack against an African-American high achiever because of his success (Asimov, 2003). This clergyman led a prayer session for the chief and command staff one Sunday. A citizen commented to the press at the prayer session that the indictment represented an attack by a white liberal district attorney against an African-American man of leadership (Morse, 2003).

Additional questions were raised about Sanders's conduct. When he accepted the job, the city attorney was investigating Sanders's involvement in a security contract proposal at the airport,

which may have involved a conflict of interest (Matier & Ross, 2002). In addition, in June 2003, a federal lawsuit was filed by two men who say they were wrongly convicted of a 1989 gang murder due to false testimony encouraged by Sanders and his partner, Hendrix. The false testimony was given by a then-14-year-old girl who later indicated that she was not at the crime scene. According to her more recent account, Sanders and Hendrix pressured her into sticking to her original story (Rosenfeld, 2003b). Allegedly, Sanders and other members of the prosecution team also suppressed a confession from an individual other than the petitioners that would have proved their innocence. Sanders had previously indicated that he did not believe the suppressed confession and suspected it was part of a gang ploy to thwart the prosecution (Rosenfeld, 2003a).

After he was indicted, Sanders stepped down from day-to-day duties, as did the six top commanders who were also indicted ("Room," 2003). However, shortly thereafter, Sanders took paid medical leave for high blood pressure, exacerbated by the stress of the indictments. Acting Assistant Chief Heather Fong took over the department (Reiterman & Glionna, 2003a), the first woman in San Francisco history to act as chief (Van Derbeken et al., 2003a). After the charges were dropped, Fagan Sr. took over the job as acting chief (Van Derbeken et al., 2003b). The mayor also stated that Sanders was welcome to resume his role as chief when his health permitted. This decision was a rebuke to the district attorney who, at the time when he dropped the charges, had specifically recommended that Sanders and Fagan Sr. not return to work, citing that they would be supervising many of the officers who may have testified against them in the grand jury (Glionna et al., 2003). Sanders resigned at the end of the year.

Chief Sanders demonstrates the pitfalls of a people-oriented police leader, based on his management of the "fajitagate" scandal involving the three off-duty police officers. Looking at the Blake and Moulton (1991) leadership grid, it is rather easy to place Sanders in the category of the country club management style. This type of leadership style is characterized by focusing more on relationships than on tasks. The assumption is that if employees are treated well, they will accomplish the tasks set before them ("being treated well" is an ambiguous concept, but in this case it meant delegating investigation to upper management). The results hardly met demands for an unbiased investigation, not surprising given the lack of supervision over a situation in which suspects included the son of the assistant chief, who was part of the upper management conducting the investigation. Sanders should have immediately removed him from handling this investigation, despite his confidence in the deputy. The task (in this case, not just conducting an unbiased investigation but also maintaining the reputation and integrity of the department) had to be of paramount importance in the decision-making process, lest the department be accused of promulgating and encouraging the infamous code of silence. Instead, relationship orientation took precedence. It might have been

a noble approach, putting trust in people ahead of the task; however, in this case, trust was not adequately allocated, and this misguided allocation tainted Sanders's otherwise impressive law enforcement career.

After leaving the San Francisco Police Department, disgraced past-chief Earl Sanders co-authored the book *The Zebra Murders* in 2006 (Russell, 2007). More recently, in 2010, the Innocence Project discovered that while still a detective, in 1994, Earl Sanders withheld a videotaped confession that would have proven the innocence of Caramad Conley who was sentenced to two life sentences without parole for murder (Muhammad, 2011). The Innocence Project also discovered that Earl Sanders paid thousands of dollars to Clifford Polk who was the main witness against Conley (Muhammad, 2011). In December 2010, a Judge held that detectives, including Sanders, knew that Polk was lying and that they purposely and knowingly orchestrated Conley's conviction (Muhammad, 2011). Conley was released from prison on January 12, 2011 (Muhammad, 2011).

CAREER HIGHLIGHTS
Task-Oriented Style

Chief Dennis Nowicki
Charlotte-Mecklenburg Police Department, North Carolina

Dennis Nowicki holds a bachelor's degree in personnel management from Northwestern University and a master's degree in public service from DePaul University (SimpliCity, n.d.). Nowicki began his police career in 1964 at the Chicago Metropolitan Police Department, rising through the police ranks to deputy superintendent of the Bureau of Administrative Services by the time he departed the force in 1989 ("Charlotte police chief," 1999). For the following three years, he served as police chief in Joliet, Illinois, which then had 230 employees ("Charlotte police chief," 1999). In 1992, he applied for the police chief

position at the Chicago Metropolitan Police Department, but ultimately he was not selected for the job (Casey, 1992). From 1992 to 1994, he was executive director of the Illinois Criminal Justice Information Authority, where he worked on developing crime-mapping technology to focus on geographic risk factors ("Charlotte police chief," 1999; Wisby, 1992). Nowicki was police chief of the Charlotte-Mecklenburg Police Department from 1994 until 1999.

In 1994, Nowicki took over the helm at the Charlotte-Mecklenburg Police Department approximately a year after Charlotte and Mecklen-burg Counties merged their police forces. He over-saw a force of 1,400 employees, which included 1,200 officers ("Charlotte police," 1994). Under No-wicki, the force operated in a relatively decentral-ized fashion, with four service areas each headed by a deputy chief. In each service area, there were three divisions led by a captain (Bostic, 1999).

According to the then City Manager Wendell White, Nowicki was selected as chief because of his experience with community-oriented policing in Joliet. Community policing was targeted as the so-lution to Charlotte's violent crime problem ("Charlotte police," 1994). In 1993, the city had 129 murders; by 1996, this rate dropped to 74.[1] Local media credited Nowicki with the reduction, citing such community-oriented programs as Right Moves for Youth, which consisted of 160 clubs that met with a police officer, a citizen, and a teacher twice a week in every school in the county ("Violent crime," 1997). Nowicki procured a $4.1 million COPS MORE (Making Officer Redeployment Effective) grant in 1997 ("Congressman Waite," 1997). By 1999, the department had 80 specifically designated neighborhood officers (Bostic, 1999).

Nowicki has been credited with introducing innovative technology to the Charlotte-Mecklenburg Police Department (Bostic, 1999). He implemented a videoconferencing system to streamline the pro-cess of acquiring misdemeanor criminal warrants. Officers save time driving to see a magistrate by being hooked up to the court system via teleconfer-ences. The system cost $65,000 and was paid for by a grant from the governor's crime commission ("New teleconferencing system," 1999). Nowicki also pro-vided each officer with a laptop computer. In addi-tion, he utilized the Geographic Information Systems (GIS) and Future Alert Contact Network (FALCON), a proactive problem alert system, in developing crime-fighting strategies in communities. For exam-ple, Nowicki studied the pattern of streetlights in the Grier Heights area in order to show that street rob-beries occurred where lighting was the dimmest. He then worked with the local electric company to im-prove lighting. Subsequently, street robberies report-edly dropped in the area (Bostic, 1999).

At the end of his tenure, Nowicki, along with Robert Guy (head of the North Carolina Department of Corrections), created Combined Enforcement and Supervision Effort (CEASE). The program consisted of police departments sharing technology, information, and police beats with area probation officers. Probation and parole officers were assigned to specific police neighbor-hood districts. The cooperation permitted police to quickly learn who was wanted for a violation of probation or parole and who was ignoring court-ordered curfews ("Charlotte leaders," 1999; "Charlotte-Mecklenburg police," 1999).

Nowicki's primary challenge as chief was maintaining a positive police–citizen relationship with Charlotte's African-American community. The fight against violent crime provoked racially charged controversies throughout North Carolina, with similar police criticism cropping up in Winston-Salem, Lumberton, Asheville, and Greensboro (Ahear & Woodall, 1994). Citizens speculated that cases with African-American vic-tims were receiving lower priority than those of Caucasian victims (Ahear & Woodall, 1994). In 1994, a city council member called for an indepen-dent investigation into the reasons for the length of time that preceded the arrest of serial killer Henry Louis Wallace, who murdered 11 African-Americna women. Ultimately, the outside audit was not undertaken. The majority of city council members agreed with Nowicki, who argued that such a review would threaten the case against Wallace. He did, however, complete an internal in-vestigation, the results of which were presented in a closed session of the city council ("Outside police probe," 1994).

A citizen community review board was cre-ated by the city council in the fall of 1997 in response to three controversial police shootings ("Family members," 1998), including the shooting of an un-armed African-American motorist, Windy Gail Thompson. Internal police investigation cleared Officer Mark Framer of wrongdoing in the shooting ("Outside police probe," 1994). The second shooting involved 19-year-old James Willie Cooper, an African-American teenager pulled over by police during a traffic stop. According to Officer Michael Marlow, who shot him, Cooper had reached into the back of his car as though he was going to pull out a gun. Police investigations cleared Marlow in the shooting. The third incident involved the shooting of a car fleeing a checkpoint. The car was shot 22 times and a shot killed passenger Carolyn Sue Boetticher, who was riding in the backseat. According to a police statement, the officers were a short distance from the roadblock interacting with another motorist when they were radioed to stop a white Chevrolet Corsica, which had been reported to be stolen. The car, in which Boetticher was traveling, swerved toward Officer Shannon Jordan. He and Officer Don Belz then opened fire as the car was speeding away (Burritt, 1997).

Nowicki's tenure continued to be plagued with controversy after the creation of the review board. In 1998, a Special Weapons and Tactical (SWAT) team drug raid led to the fatal shooting of Charles Irwin Potts. According to police, Potts brandished a gun at SWAT team member James D. Guard, who then shot him. Guard was subsequently placed on administrative duty pending an internal investigation. At the time of the drug raid, 10 people were in the house and 2 of them contradicted police, telling the press that Potts did not point a gun before he was killed. Press reports quoted Nowicki: "The Charlotte-Mecklenburg Police Department is well respected for its thorough and objective investigations of police use of force. There aren't any reasons to mistrust the police department" ("Family members," 1998).

In 1999, Nowicki left policing to become a law enforcement consultant and head of the Center for Public Service and Leadership at Pfeiffer University in Charlotte ("Deputy chief," 1999). At the center, Nowicki has continued developing crime-mapping technology. Along with the Carolina Institute of Community Policing, Nowicki trained neighborhood leaders in how to use GIS to track and fight crime ("The state/police," 1999). In 2001, he was a consultant to the National Center for Children Exposed to Violence (Horn & Prendergast, 2001). Nowicki has also been a consultant to the U.S. Department of Justice Civil Rights Commission,

Yale University Child Study Center, and the National Institute of Justice (SimpliCity, n.d.).

Unlike Sanders (whose career was marked by one defining incident), Nowicki faced multiple challenges throughout his career in the Charlotte-Mecklenburg Police Department. His leadership style is very consistent with task orientation, particularly in the way he initiated the structure approach in response to these challenges.

Nowicki appears to exhibit the participative style of leadership, one that emphasizes the use of group supervision rather than the supervision of each individual employee. Although the joint decision-making aspect of participative leadership can be divided into four main categories, his style seems to match specifically the delegation category, whereby the leader gives the group authority to make decisions, delineating the outer limits of their choices. The author participated in one of Nowicki's briefings of his command staff during his tenure as chief, and she vividly recalls the chief addressing his deputies on a certain issue of concern and finishing the discussion by stating the following: "I am a big believer in delegation and empowerment, but I am the one who is going to define the outer perimeter of this empowerment." It was a very telling and powerful statement that is backed not just by the words spoken during this briefing but also by the events described in his biography.

Indicators of this task orientation/participative style include implementation of community policing ideas, responses to various use-of-force incidents that plagued his tenure, decentralization of the force into four service areas and empowerment of the deputy chiefs to lead these areas (if closely supervised), and many innovations and technology-driven initiatives that emphasized cooperation and high accountability. His approach was not completely devoid of relationship orientation, of course, but the task to be accomplished remained the focus of his efforts. It is difficult to imagine a scandal of "fajitagate" proportions ever occurring in Charlotte with these safeguards and priorities.

Since leaving the Charlotte-Mecklenburg Police Department, Dennis Nowicki has worked as a criminal justice consultant for numerous agencies including the Department of Justice (Office of the Independent Monitor for the Metropolitan District of Columbia Police Department, 2007). He also contributed to a 2005 report on the Metropolitan Police Department and a 2010 report on the Virgin Island Police Department (Office of the Independent Monitor, 2005; Office of the Independent Monitor, 2010).

[1]This same year violent crimes dropped by 7 percent nationwide, with an 11-percent drop in murders ("Violent crime," 1997).

Exercise

The police chiefs detailed in this chapter have been examined through the prism of style theory. Using what you have learned from other theories in this book or from the summaries (as indicated in Introduction) you have read thus far, it is possible to consider their approaches through the lens of different theories. What other theories can you find applicable to these chiefs? Are there lessons to be learned from those applications, or do those theories indicate alternative paths or approaches that might have yielded a more successful outcome?

References

Ahear, L. & B. Woodall (1994, June 26). Civilian boards: Mixed reviews. *News & Record*, p. A3.

An officer for justice. (2002, July 15). *The San Francisco Chronicle*, p. B4.

Asimov, N. (2003, March 1). Race factor evokes complex responses from blacks: Center backs Sanders, but watchdogs wary of blue uniforms. *The San Francisco Chronicle*, p. A18.

Bass, B.M. (1990). *Bass and Stogdill's handbook of leadership: A survey of theory and research*. New York: Free Press.

Blake, R.R. & A.A. McCanse (1991). *Leadership dilemmas—Grid solutions*. Houston, TX: Gulf Publishing.

_____ & J. Moulton (1982). The uncertain future of the leadership concept: Revisions and clarifications. *Journal of Applied Behavioral Science*, 18 (3), 275–291.

Bostic, H., Jr. (1999, March 3). Meeting report. Retrieved May 21, 2003, from www.charlotterotary.org/march2_99.htm.

Bowers, D.G. & S.E. Seashore (1966). Predicting organization effectiveness with a four-factor theory of leadership. *Administrative Science Quarterly*, 11, 238–263.

Burritt, C. (1997, April 10). Charlotte police defend roadblock shooting. *The Atlanta Constitution*, p. 6A.

Cartwright, D. & A. Zander (1960). *Group dynamics research and theory*. Evanston, IL: Row, Peterson.

Casey, J. (1992, February 3). City gets 40 applicants for top police post. *Chicago Sun-Times*, p. 5.

Charges against San Francisco police chief dropped. (2003, March 12). *CNN.com*. Retrieved July 23, 2003, from http://cnn.com.

Charlotte leaders praise CEASE initiative. (1999, May). *North Carolina Department of Corrections News*. Retrieved May 25, 2003, from http://www.doc.state.nc.us/NEWS/1999/9905news/CEASE.htm.

Charlotte-Mecklenburg police to track 8,000 criminals on probation, parole. (1999, April 5). *The Associated Press*, state and local wire.

Charlotte police chief announces resignation. (1999, February 15). *The Associated Press*, state and local wire.

Charlotte police recruit Chicago veteran as chief (1994, March 8). *News & Record*, p. B2.

Chief Sanders deserves better. (2003, March 11). *The San Francisco Examiner*, p. 12.

Chronicle of events leading up to SF police indictments. (2003, March 9). *The Associated Press*, state and local wire.

Congressman Waite announces $4 million COPS MORE grant to Charlotte. (1997, January 29). Press release. Retrieved May 25, 2003, from http://www.house.gov/watt/pr970129.htm.

Dainty, P. (1986). Leadership and leadership research. *Journal of Managerial Psychology*, 1 (1), 40–43.

Deputy chief among candidates for Charlotte-Mecklenburg police. (1999, August 7). *The Associated Press*, state and local wire.

Family members of slain man vow to take case to review board. (1998, September 7). *The Associated Press*, state and local wire.

Glionna, J.M. & T. Reiterman (2003, March 10). S.F. mayor angry about scandal: Willie Brown says he blames himself as well as others for the police controversy. *Los Angeles Times*, Metro, p.1.

_____, T. Reiterman & M. Dolan (2003, March 12). Case against chief, top aide dropped. *Los Angeles Times*, p. 1.

Horn, D. & J. Prendergast (2001, May 15). Justice Dept. to begin police review. *The Cincinnati Enquirer*. Retrieved May 25, 2003, from http://www.enquirer.com/editions/2001/05/15/loc_justice_dept_to.html.

Kahn, R.L. (1956). The prediction of productivity. *Journal of Social Issues*, 12, 41–49.

Likert, R. (1961). New patterns of management. New York: McGraw-Hill.

Matier, P. & A. Ross (2002, July 14). S.F. top cop: The story behind the badge. *The San Francisco Chronicle*, p. A21.

Misumi, J. (1985). The behavioral science of leadership: An interdisciplinary Japanese research program. Ann Arbor, MI: University of Michigan Press.

Morse, R. (2003, March 3). S.F.'s indicted police chief joins in prayer for justice. *The San Francisco Chronicle*, p. A10.

Muhammad, C. (2011, May 14). Lessons on life through wrongful conviction. *The Final Call*. Retrieved May 31, 2011, from http://www.finalcall.com/artman/publish/National_News_2/article_7834.shtml.

New teleconferencing system saves officer time. (1999, February 26). *The Associated Press*, state and local wire.

Northouse, P.G. (2004). Leadership: Theory and practice (3rd ed.). Thousand Oaks, CA: Sage.

Outside police probe vetoed in Charlotte: Serial killer investigation. (1994, May 10). *News & Record*, p. B2.

Office of the Independent Monitor. (2005). *Eleventh quarterly report of the Independent Monitor for the Metropolitan Police Department*. Retrieved from Clearing House website: http://www.clearinghouse.net/chDocs/public/PN-DC-0001-0015.pdf.

_____(2010). *First quarterly report of the independent monitor for the Virgin Islands Police Department*. Retrieved from Office of the Independent Monitor website: http://www.vipd.gov.vi/Libraries/Non-Form_PDFs/100507report.sflb.ashx.

Office of the Independent Monitor for the Metropolitan District of Columbia Police Department. (2007). *About the office of the independent monitor*. Retrieved from Office of the Independent Monitor for the Metropolitan District of Columbia Police Department website: http://www.policemonitor.org/MPD/about.html.

Police scandal rocks San Francisco. (2003, March 4). *CBSNews.com*. Retrieved July 23, 2003, from http://www.cbsnews.com/stories/2003/03/04/national/printable542678.html.

Reddin, W.J. (1967). The 3-D management style theory. *Training and Development Journal*, 21 (4), 8–18.

Reiterman, T. & J.M. Glionna (2003a, March 5). S.F. enters plea of innocent: He and six aides are arraigned in alleged cover-up. *Los Angeles Times*, Metro, p. 1.

_____(2003b, March 6). S.F. chief says D.A. was invited to direct probe: In a brief filed with state attorney general,

police leader says he met prosecutor twice. *Los Angeles Times*, Metro, p. 1.

_____(2003c, March 13). S.F. police chief recounts "the worst day" of his life. *Los Angeles Times*, Metro, p. 1.

Room for leadership. (2003, March 4). *The San Francisco Chronicle*, p. A20.

Rosenfeld, S. (2003a, April 6). SFPD in crisis: New bombshell in old homicide. *The San Francisco Chronicle*, p. A1.

_____(2003b, June 19). Witness: DA, cops, coached me to lie; Sanders, partner accused in tainted murder testimony. *The San Francisco Chronicle*, p. A1.

Russell, R. (2007, February 21). Earl's last laugh. *San Francisco Weekly*. Retrieved May 31, 2011, from http://www.sfweekly.com/2007-02-21/news/earl-s-last-laugh/.

San Francisco police chief, others indicted in connection with fight. *CNN.com*. Retrieved July 23, 2003, from http://cnn.com.

S.F.'s new chief wants to be fully staffed (2002, July 20). *The San Francisco Chronicle*, p. A16.

SF police chief asks state to investigate his indictment. (2003, March 1). *MSNBC.com*. Retrieved July 23, 2003, from http://stacks.msnbc.com/local/kntv/a1514201.asp.

SimpliCity. (n.d.). Simply connecting citizens to government. Retrieved May 21, 2003, from www.simplicity311.com/team.htm.

The state/police, area leaders can learn system: University offers latest anti-crime technology. (1999, December 14). *Wilmington Star-News*, p. 3B.

Stogdill, R.M. (1974). *Handbook of leadership: A survey of theory and research*. New York: Free Press.

Sward, S. & B. Wallace (2003, March 2). SFPD's dismal record battling misconduct: Lawsuits, complaints, jury awards no bar to promotions. *The San Francisco Chronicle*, p. A1.

Van Derbeken, J. (2002, August 11). Homicide shakeup at SFPD: New chief changes head of troubled unit. *The San Francisco Chronicle*, p. A1.

_____ (2003a, January 23). Chief defends handling of beating case: SFPD smeared by media, he tells police commission. *The San Francisco Chronicle*, p. A15.

_____ (2003b, July 24). D.A. to refile in fajita case. *The San Francisco Chronicle*, p. A19.

_____& S. Sward (2002a, November 23). Mayor, chief soft-pedal cop fracas. *The San Francisco Chronicle*, p. A1.

_____ & S. Sward (2002b, December 12). D.A. sees holes in probe of cops. *The San Francisco Chronicle*, p. A25.

_____, R. Gordon & J.H. Zamora (2003a, March 4). S.F. police chief, aides step aside: Sanders on sick leave; others suspended. *The San Francisco Chronicle*, p. A1.

_____, R. Gordon & J.H. Zamora (2003b, March 12). Fagan put in charge: He'll be acting chief mayor says. D.A. drops case against top cops. *The San Francisco Chronicle*, p. A1.

Violent crime in U.S. falls a record 7%, local results mixed. (1997). *The Virginian-Pilot*, p. A3.

Wisby, G. (1992, September 10). Plan fights gangs, domestic violence. *Chicago Sun-Times*, p. 11.

Yukl, G. (1994). *Leadership in organizations* (3rd ed.). Englewood Cliffs, NJ: Prentice Hall.

Zamora, J.H. & I. Lelchuk (2002, July 12). New chief paid his dues for decades. *The San Francisco Chronicle*, p. A17.

Zelditch, M. (1955). Role differentiation in the nuclear family: A comparative study. In T. Parsons, R.E. Bales and J. Olds (Eds.), *Family socialization and interaction process* (pp. 307–351). Glencoe, IL: Free Press.

When the Event Is Just Too Much to Handle
Situational Leadership Theory

OVERVIEW OF THE THEORY

Situational leadership theory focuses on what leaders should do given the nature of their subordinates. In this chapter, Montgomery County Chief Charles A. Moose—who handled the high-profile sniper shootings in suburban Washington, D.C., in 2002—illustrates the complexity of managing a diverse set of subordinates. LAPD Chief Bernard Parks (whose approach to a LAPD already riddled with allegations of misconduct then backfired when yet another scandal, Rampart, was exposed) demonstrates the other case study. This theory expects leaders to vary their approach depending on their subordinates. In essence, the level of maturity of the subordinates dictates the best leadership style, building on the 3-D leadership framework developed by Reddin (1967). The theory, however, adds an effectiveness dimension (Hersey & Blanchard, 1969). A study by Stogdill (1974) supports the notion that "the view that leaders tend to change certain aspects of their behavior in response to changes in group task demands" (p. 16). Bass (1990) explains:

> It is apparent that the kind of leader who emerges in an organization and the individual who is successful as a leader and is evaluated as effective as a leader depends on the philosophy of the larger organization in which the leader's group is embedded. (p. 571)

Subordinate maturity is the factor that varies and dictates the best leadership style (consideration versus structure, or directive behaviors versus supportive behaviors) in a particular situation. It is measured in regard to the specific task to be accomplished by the subordinate. There are two components to maturity: job maturity and psychological maturity (in a later model, it is referred to as "willingness"). The first refers to the subordinate's ability; the latter refers to the subordinate's self-confidence related to the task (Yukl, 1994). Both are composite variables of what Hersey and Blanchard (1969) take to be the main factors involved in maturity (Yukl, 1994). For example, job maturity focuses on the ability the subordinate has, which is a function of knowledge, experience, and skill. Willingness is operationalized as the extent to which the employee has confidence, commitment, and motivation for the work (Fernandez & Vecchio, 1997; Northouse, 2004).

Originally conceptualized as life cycle theory, it was later modified and renamed as situational leadership by Hersey and Blanchard (1969). They start with the basic premise that there is no normative or best style of leadership; instead, successful leaders are flexible enough to adapt and accommodate to their followers (Hersey & Blanchard, 1969). Hersey and Blanchard (1969) thought of the workplace structure as teams in which leaders exist to help followers accomplish organizational goals. Implied by this are duties such as performance planning, day-to-day coaching and counseling, and performance evaluations based on measurable objective criteria. Goals are determined based on the criteria suggested by the acronym SMART (specific, measurable, attainable, relevant, and trackable; Blanchard et al., 1985).

Moreover, underpinning the theory is the notion that organizations are open systems, meaning that what takes place outside a system affects what happens inside the system (Bass, 1990). Subordinate maturity, therefore, is a function of preceding experiences of subordinates that shape both their ability to complete task objectives and their psychological maturity.

Empirically, there seems to be more support for the notion that job maturity, rather than willingness or psychological maturity, is the critical follower attribute (Fernandez & Vecchio, 1997). However, according to researchers, there are three problems with this theory: 1) maturity is operationalized vaguely; 2) the Leadership Effectiveness and Adaptability Description (LEAD) is biased toward one category; and 3) the validity of the tenets of this theory are weak (Dainty, 1986; Graeff, 1983, 1997; Yukl, 1994). Despite numerous criticisms, however, the theory survives primarily because it has strong intuitive appeal; it has also benefited from being a management fad in the 1980s (Graeff, 1997).

Chadbourne (1980) uses the notion of subordinate maturity to determine trainers' leadership styles when teaching trainees. She describes the organic development of a training group. In the beginning, "the leader takes initial responsibility for building a conducive learning climate through structuring, modeling effective behaviors and providing didactic input" (p. 55). In the middle stages of the training, the leader encourages team building through small-group activities (with less structure flowing directly from the leader), which provide time for reflection on the material to be learned. Finally, as the trainees gain job and psychological maturity related to the training material, the leader has effectively provided the trainees with the opportunity to pursue their goals and personal growth on their own.

In addition, the opposite is possible. Leaders' behavior can hypothetically influence the subordinates' level of maturity. According to Hersey and Blanchard (1969), they can utilize developmental interventions in order to create the level of maturity desired. In the event that the leader and subordinate disagree as to the subordinate's level of maturity, the leader may need to concede to the subordinate's view in order to maintain the task environment. As Bryman (1986) speculates: "This possibility forces one to question how far the maturity level of subordinates is a deterministic model, if the leader may have to adopt what he believes to be an inappropriate style in terms of his own assessment" (p. 148).

Contingency contracting, a developmental intervention, involves a negotiated agreement between the leader and the subordinate about the tasks and objectives and the leader's role in accomplishing the desired outcomes. By creating such a predetermined plan of attack the subordinate can reach a higher maturity level (Yukl, 1994).

According to Blanchard and Nelson (1996), the best managers not only provide the direction and support commensurate with maturity but also recognize employees for their efforts. This can assist the leader in communicating that the achievement of high maturity and task effectiveness is not without reward. However, regardless of subordinates' stage of maturity, each stage has a corresponding best practice for providing recognition: S1 types (enthusiastic beginners with low maturity) need the attention of specific direction and redirection on a daily basis; S2 types (disillusioned learners) need feedback on their performance and face-to-face praise; S3 types (capable but cautious learners) need public praise and perhaps an award such as a photo on a "wall of fame"; and S4 types (self-reliant achievers with high maturity) require similar public rewards as the S3 but can also be rewarded by being given the task of training other, less mature workers.

SITUATIONAL LEADERSHIP THEORY AND EMPLOYEE BURNOUT

Niehouse (1984) states that the theory has particular applicability when confronting employee burnout, the symptoms of which include chronic fatigue, job boredom, cynicism, lack of recognition, less commitment, moodiness, poor concentration, and physiological changes. Niehouse (1984) argues that leadership decisions can affect employees negatively when not applied in a way that is appropriate both to the situation and to the subordinates' level of maturity, creating a fight-or-flight syndrome which then leads to burnout. Borrowing from situational leadership theory, Niehouse (1984) states that, for example, burnout can occur easily when a leader provides too much structure to a mature capable employee. However, besides appropriate style, Niehouse (1984) suggests that any leadership changes should be implemented gradually, without ambiguity, and with realistic goals in mind.

SIMPLIFIED SITUATIONAL LEADERSHIP THEORY

Van Auken (1992) proposes a simplified version of situational leadership theory. He puts forth four main leadership situations and the styles they demand:

1. *The political arena.* Involves issues where the employees have a stake in the outcomes, although they are not the substantive issues of the work at hand. Here, the supervisor functions with a "politician" style in trying to keep "constituents" happy.
2. *The bureaucratic arena.* Involves routine operations requiring decisions best made without subordinate involvement in order not to needlessly waste subordinates' time.
3. *The technical arena.* Technical problems are an opportunity to solicit subordinate involvement in problem solving in an informal manner. Even if the problem has a minor effect on employees, they are flattered to be asked to weigh in on the issue.
4. *The professional arena.* Participative management is ideal when subordinates are affected by an issue and have the skills to handle it. Setting goals and creating deadlines are examples of issues in which the leader would be foolhardy not to involve employees. Consensus building on key issues keeps subordinates happy and functional.

POLITICS AND SITUATIONAL LEADERSHIP THEORY

In the political realm, leadership style exists along a continuum from centralized authoritarian leadership to participative democracy. The leadership question then becomes, under what societal conditions leaders should adopt one of the two styles? According to Korten (1962), the situational moderator variables are either high goal structure or low goal structure. In high goal structure, groups in society have specific goals that they have determined via consensus. In low goal structure, groups have fewer shared objectives; the emphasis is on individual rather than shared goals. From the above, Korten (1962) envisions a four-cell matrix with structural variables on one axis and leadership styles on the other. Intervening is also a stress variable. A group's perception of

Individuals vs. Group Subsystems
- What happens is not as important as how you react to what happens.
- Police leaders are constantly confronted with conflict. How they manage it is their litmus test.

Two dimensions to manage
- Internal
- External

immediate threats, and the anxiety or stress caused by such threats, determines shifts in the style. Political leadership that is particularly autocratic is more appropriate in situations in which there are stressful ambiguities, in which immediate reaction is deemed necessary, and in which there is consensus on society-wide goals. However, authoritarian leadership cannot maintain itself without stress. When the stress or crisis subsides, individual needs will be the priority, and participative democracy can better satisfy such needs. This political or crisis theory dovetails with aspects of this theory that might be most applicable to policing, which were discussed above.

POLICING AND SITUATIONAL LEADERSHIP THEORY

Hersey and Blanchard (1969) indicate that a high task style is the most appropriate for crisis-oriented organizations. This is because success in stressful situations often depends on unequivocal responses to orders from leaders. Therefore, it is likely that Hersey and Blanchard would apply the high task style to police agencies. Bass (1990) similarly argues that unstable environments lead to organizational policies that are less uniform. Leadership styles will change to be flexible and to respond to the ever-changing reality. Crime-control agencies can also be said to operate in a world of shifting crime trends and crime-control strategies.

Consistent with the above and drawing from Hersey and Blanchard (1977), Kuykendall and Unsinger (1982) surveyed police supervisors and found that the most common management style was one that was high task and high relationship, as defined under style leadership (see Chapter 6)

Types of Conflict	**Conflict Factors**
Intrapersonal	Power
Interpersonal	Control
Structural	Frustration
Strategic	
Community	

Conflict as a Dynamic Process
- *Latent conflict.* Condition exists but is not recognized.
- *Perceived conflict.* Condition is recognized by both parties.
- *Felt conflict.* Internal tensions exist but are not apparent.
- *Manifest conflict.* Conflict is open and obvious to involved parties.
- *Conflict aftermath.* New conflict or cooperation results.

and situational leadership theories. The other frameworks that respondents could select from included "Great Man" (trait theory; see Chapter 12), environmental (time, place, and circumstance), exchange (of mutual rewards), humanistic (intersection of individual and organizational interests), and interaction-expectation (ability to interact to get a desired response in subordinates).

In a 1982 study, Kuykendall and Unsinger administered Hersey and Blanchard's LEAD instrument to 155 police managers in training programs. The instrument measures managerial style range and adaptability; style range refers to the number of styles a leader uses, and adaptability refers to the ability of the manager to fit the style to the situation. Police managers were given the instrument at the onset of the training program. The managers were asked to respond to the scenarios described, using the available task- or relationship-oriented styles, as they would in the workplace. There were four styles to choose from:

1. *Telling.*   This refers to high task, low relationship orientations.
2. *Selling.*   This is characterized by high task and high relationship approaches.
3. *Participating.*   This encompasses high relationship, low task orientations.
4. *Delegating.*   This reflects low relationship, low task approaches.

The majority of participants had 5+ years experience, had some college education, and came from departments with less than 100 officers. In addition, almost half had at least four years of college education.

Most managers indicated that they either had no dominant style or utilized a selling approach (high task, high relationship). None of the managers used all four possible styles, but 97% used two or three styles. The instrument also provides effectiveness scores. The scores reflected that 80% of the managers were somewhat to very effective. The instrument also measures to what degree the police managers were able to match the style to the situation. Results reflected that managers were most effective with a high task emphasis (selling and telling) and were least effective with low task approaches (participating and delegating). Delegating was the style that was rarely used (102 managers did not use it at all, and 31 managers used it only once). The analysis further showed that 78% of police managers tended to use participating/telling and telling/selling combinations. Selling was the most common management style employed. Hersey and Blanchard (1977) consider the selling style a safe approach because it is an effective style in most situations, even when it is not the most effective option.

Hersey and Blanchard (1977) posit that the lack of delegation style may be unique to police environments. Delegation may counter the occupational emphasis on action over inaction because delegation is inactive. The factors that may contribute to the action imperative are 1) a self-perception about the role of policing, 2) the reform model of police thinking that dominated the 20th century, 3) a perception of behavior that is desirable under stressful situations, and 4) the consequences of "looking bad" in police work.

> Admittedly, the variety of situations in which police can become involved are, perhaps, unlimited. And the range of possible responses considered effective may be a function of the specific problem and its situational context. This, of course, means that providing definitive guidelines for what to do is extremely difficult. However, this often leaves the police in the position of being told they were wrong after a decision without previous guidance as to what was considered "right." Lacking such guidance, some police officers tend to "play it safe" while police managers come to believe that they must monitor the decisions made by their subordinates. (Kuykendall & Unsiner, p. 320)

Situational leadership continues to be a model advocated in police environments. The Community Policing Consortium (1997) presents situational leadership as the theoretical model underpinning the strategies it teaches as part of its training for police supervisors. Its advocacy of the model is based on its flexibility in getting others to act in accordance with organizational values. Citing Hersey (1985), the Community Policing Consortium (1997) provides training that encourages police supervisors to diagnose a police officer's ability and willingness to perform a task (the police officers readiness, indicated by observable skills, experience, and training) and to assess the best level of relationship and task behavior to apply in that situation.

SITUATIONAL LEADERSHIP THEORY AND THE CASES OF CHIEF CHARLES A. MOOSE AND CHIEF BERNARD PARKS

Situational leadership theory emphasizes the importance of a given situation or event that a leader faces. The way the leader chooses to behave is directly dependent on the assessment of the subordinates and their level of commitment and competence. The level of intervention within any given situation will depend on the level of willingness and readiness of the police officers involved.

The two chiefs identified for this chapter represent two different approaches to the situational leadership style. Chief Charles A. Moose exemplifies the style 1 (S1) level of leadership, exhibiting above-average orientation toward the task while little time is allocated to the supportive mechanisms. When achievement of the task becomes the primary orientation, subordinates can be prevented from providing input and their feelings are minimized. Chief Bernard Parks, on the other end, displayed the other extreme, the style 4 (S4) level of leadership. In this approach, also known as the delegating style, the leader is convinced that his/her involvement in both task and relationship can be minimized and in essence gives control over to the subordinates. The point here is to illustrate the theory and should by no means be interpreted to mean that in all situations the chiefs used these methods.

Overview of Situational Leadership Theories

Situational leadership refers to the interaction between these factors:

- Amount of guidance and direction (task behavior) a leader gives
- Amount of socioemotional support (relationship behavior) a leader provides
- Readiness level that followers exhibit in performing a specific task, function, or objective

Four styles of situational leadership:

Style 1 (S1). Characterized by above-average amounts of task behavior and below-average amounts of relationship behavior

Style 2 (S2). Characterized by above-average amounts of both task and relationship behaviors

Style 3 (S3). Characterized by above-average amounts of relationship behavior and below-average amounts of task behavior

Style 4 (S4). Characterized by below-average amounts of both relationship and task behaviors

Effectiveness of each style is dependent on the given situation.

(Developed by Hersey and Blanchard, 1969 [later refined by Hersey and Blanchard, 1977, 1988, 1993; Blanchard et al., 1985, 1993])

CAREER HIGHLIGHTS
Situational Leadership Style (S1)

Chief Charles A. Moose
Police Bureau, Portland, Oregon; and County Police Department, Montgomery, Maryland

Charles A. Moose was a celebrated and high-profile police chief even before he oversaw the investigation of sniper shootings in suburban Washington, D.C., in the fall of 2002.

Moose was born in 1953, New York City, while his father was attending Columbia University. The family left the city for Lexington, North Carolina, when he was a toddler. Moose, an African American, attended segregated schools through the sixth grade. He attended the University of North Carolina, Chapel Hill, earning a degree in U.S. History. While in college, he intended to go to law school, but a professor encouraged him to take a police exam to earn extra credit. During his senior year, he was offered a job as a police officer in Portland, Oregon, which he accepted. He intended to police for a few years before turning to law school but has remained in policing ever since (Fermino & Kranes, 2002).

Moose joined the Portland Police Bureau in 1975 (Bernstein, 1999a). While an officer, he attended the 154th session of the FBI National Academy in 1988 ("Vital Stats," n.d.). He later earned a master's degree in public administration in 1984 and a Ph.D. in urban studies and criminology in 1993 from Portland State University. He has one academic publication in a peer-reviewed journal about the impact of community-policing values on the implementation of a domestic violence reduction strategy in Portland, Oregon, which appeared in *Crime and Delinquency* in 1997. He was on the adjunct faculty in criminal justice at Portland State University from 1994 to 1995. From 2000 until the present, he has been on the adjunct faculty at Montgomery College ("Vital Stats," n.d.).

Racial issues have always been a point of tension for Moose. Between 1975 and 1992, he was disciplined four times for excessive anger at people he thought where discriminating against him while on the job (Bernstein, 1999c; Fermino & Kranes, 2002). In 1992, he blew up at a retail clerk he thought was not serving him and shouted, "Does my skin have to turn white to get help?" (Sisk, 2002).

Chief of the Portland Police Bureau (1993–1999)

Moose served as chief of the Portland Police Bureau from 1993 to 1999. Among controversies in Moose's tenure as police chief in Portland was his decision to use shotguns loaded with beanbags to control crowds. The beanbag rounds are made of nylon sacks filled with lead shot that are designed to injure but not kill. The issue was brought to the fore on August 17, 1998, when approximately 40 people, protesting the closure of a park in the King neighborhood of the city (where Moose also lived), protested in front of his house. Police used the beanbag ammunition against the crowd. Moose justified the decision later by saying that the ammunition was more safe and effective than other options. He also stated in a public forum that he and his wife had made the wrong decision to live in the King section of Portland, an area with an African-American majority and a high crime rate (Greimel, 1998). Originally, in late 1993, he indicated that he bought the house in King to signal that he was committed to rebuilding the neighborhood and to promote community policing. He believed that community-policing ideals included the concept that officers should live in the communities they protect (Nkrumah, 1999).

In January 1999, Moose took over the school police in Portland, which had been operating independently of the police force. The impetus for the change was cutting administrative costs of policing schools by the city. The decision was unpopular among school officers who were concerned about internal upheaval and career paths ("Plan," 1999).

In a popular move, Moose and Portland-area police chiefs signed a joint resolution in April 1999 denouncing racial profiling by police conducting traffic stops. According to the U.S. Department of Justice (DOJ), it was the first such resolution by law enforcement officials in the nation. The resolution, drafted by Moose, was a response to increased criticism that police target minorities disproportionately in automobile stops. Contemporaneously, Attorney General Janet Reno had called for increased data collection to identify the scope of the problem. Moose, in statements to the press, appeared ready to be accountable to the public for any future such profiling when he said, "If we're perceived to be engaging in this behavior, I expect to get this document back in my face" (Oshiro, 1999, p. A1).

The chief had been praised throughout his tenure for developing community policing (Bernstein, 1999a), receiving $26.6 million in federal and state money, some of which was used to hire 60 additional officers. For example, he had regular meetings with community leaders, called "Chief's Forums" (Bernstein, 1999c). He also raised the bar on the educational background required of new recruits to a four-year college degree (Bernstein, 1999a). Officers praised Moose for choosing to wear his uniform and not losing his connection to the street (Bernstein, 1999c).

However, he received criticism for his occasional angry outbursts (Bernstein, 1999a). One media source reported that he was ordered to take anger management classes while serving as Portland's police chief (Duin, 2002). His style of management was described by the media as "heavy-handed" (Bernstein, 1999c). Charles Makinney, who headed the bureau's fiscal and records division from 1981 to 1994, told reporters, "[H]is ability to deal as a manager was horrendous. If he was able to deal with personnel better, I think he would have been a better chief" (Bernstein, 1999c, p. A1). The subsequent fiscal and records manager, Nancy Dunford, who

resigned in 1997 after clashing with Moose, said, "He wasn't able to generate community spirit in his own bureau. People were attacked. It was dictatorial, not participatory" (Bernstein, 1999c, p. A1). During the city's investigation of the use of beanbag-style ammunition, Moose yelled at the mayor and city commissioners, saying that the inquiry was a waste of time (Barnett, 2002; Bernstein, 1999c).

Moose was also criticized for his handling of internal police investigations (Bernstein, 1999a). In 1996, he investigated a central precinct commander over allegations that the commander routinely hired male prostitutes. Although the commander was not indicted, Moose sought to terminate his career. The commander then filed a federal lawsuit against Moose, claiming he was investigated because he was gay. In addition, in April 1997, Moose disciplined two commanders for using their work-issue cell phones for personal calls; then he changed his mind and returned them to work without discipline. He subsequently disallowed all officers' use of city cell phones, angering officers and the community (Bernstein, 1999c).

On September 13, 1997, David Moose, the chief's then-17-year-old son (by his first wife), and his 16-year-old friend were arrested for possession and distribution of an undisclosed amount of crack cocaine. He was booked in juvenile detention but released the same day on bail. According to news reports, Moose praised his officers for making the arrest even though they were aware that the suspect was his son ("Son," 1997). The charges were subsequently dismissed in court because the drugs were discovered pursuant to an illegal search and seizure during a stop for jaywalking (Video Monitoring Service, 1997).

Upon his departure from Portland to take the police chief job in Maryland, the governor declared June 28, 1999, Chief Charles A. Moose Day for his leadership in law enforcement and his commitment to community-oriented policing. The Oregon Association of Chiefs of Police gave Moose its President's Award for promoting a diverse police force and for reaching out to the community (Bernstein, 1999b).

Police Chief of Montgomery County (1999–2003)

Moose had unsuccessfully sought a position as police chief in Washington, D.C. (Bernstein, 1999a). He took over the Montgomery County Police in suburban Rockville, Maryland, in August 1999, after Montgomery County Executive Douglas Duncan stated that Moose was "a natural leader" and recruited him (Sisk, 2002). Upon instatement, Moose indicated that he planned to confront racial issues. When his tenure began, the U.S. Department of Justice was conducting a review of deadly use of force in the county after several police shootings involving minorities (Holly, 1999).

In the fall of 2002, two snipers, John Lee Malvo and John Mohammad, held the public in fear for weeks as they carried out sniping attacks against random civilians. Moose headed the high-profile investigation involving thousands of law enforcement officers from local, state, and federal agencies. He was hailed by the public after the capture of the two snipers. President George Bush was compelled at one point during the investigation to say that the FBI would not be taking over the investigation for Moose, as they had hoped to do (Sisk, 2002). After the investigation, media coverage suggested that Moose had become "America's policeman" (Kennedy & Becker, 2002). As one source reported, "He was unprepared, unrehearsed and seemingly incapable of staying calm in the face of the growing carnage. . . . Moose then was the perfect metaphor for [the public's sense of] helplessness" (Duin, 2002, p. A1). On November 24, 2002, people wore moose antlers made of purple foam in his honor in a parade ("Parade," 2002). Subsequently, Moose was in high demand as a public speaker (Manning, 2003). He spoke at such venues as the American Leadership Forum of Oregon (Tsao, 2003); he was also featured in a national television interview by Barbara Walters (Terzian, 2003).

Despite the successful outcome in the sniper investigation, Moose was criticized for his handling of the case (Sisk, 2002). One of the criticisms revolved around the information he gave the public, which some have said incited fear rather than reassurance. For example, early on in the crisis he stated that the elusive sniper "may be gloating" nearby (Clines, 2002, p. A12). Furthermore, the police union formed a special committee to investigate his handling of lookout information during the manhunt. Officials believe that Moose endangered patrol officers by withholding critical information, in violation of a recently signed labor contract in which Moose agreed to a safety provision involving the sharing of information with officers. In April 2003, testimony was heard from detectives who worked the sniper manhunt that indicated Moose may have held back information about the snipers' physical descriptions. There was also evidence that he had identified the suspects as early as October 22, 2002 (Sperry, 2003). However, Moose continued to allow the police and the public to search for unidentified white suspects in a white van instead of the allegedly then-identified black suspects in a blue Chevrolet Caprice (Malkin, 2003; Washington, 2002).

In addition, Moose's volatile temperament, cited throughout his career, returned, this time in the public spotlight. He cried in front of the media while reporting the sniper shooting of a 13-year-old. He also lashed out at the media for reporting about a note, "I am God," left by the sniper on a tarot card depicting the skeletal figure of Death (Barnett, 2002). In yet another emotional display, while debasing former cops and FBI profilers who had speculated about the killer on television, he threatened to turn the case over to the media: "[A]nd you can solve it" (Kennedy, 2002, p. 7).

In January 2003, *The Washington Post* reported that Moose signed book and movie deals based on his leadership role in the sniper investigation ("Top sniper cop," 2003; Tsao, 2003). One deal was made with E.P. Dutton publishers in which Moose was to write a first-person account of the sniper investigations. The second deal was for a television movie with a Hollywood production company. In addition, after the sniper investigation, he and his wife, Sandy Moose, started a crisis-management consulting company, which has not yet been profitable (Terzian, 2003). The deals and consulting company triggered a Montgomery

County Ethics Commission hearing. According to the relevant county ethics law, a public employee is not permitted to intentionally use the prestige of a public office for private gain (Manning, 2003).

In late March 2003, the ethics panel ruled unanimously that Moose could not write a book or consult for a movie based on the sniper suspects while also serving as the county's police chief. His biggest supporter, County Executive Douglas Duncan, lobbied for the ethics panel to give him a waiver or to create legislation with an exception to the rule (Terzian, 2003). Losing the battle, Moose resigned from the force at the end of June 2003; in September 2003, he released his book about his career and the sniper investigations: *Three Weeks in October: The Manhunt for the Serial Sniper* ("Charles Moose," 2003; "Statement," 2003).

Throughout his career, Chief Moose has displayed a very low level of control over his emotional intelligence (EQ), for example, by shedding tears on national television and having characteristic angry outbursts. During times of crisis and fear, a chief of police must display to the public strength and reassurance, qualities which demand complete control over EQ. If a chief breaks down, then the message sent to the general public is vulnerability, the last emotion a person in a position of leadership should convey or want to convey.

However, this emotionalism was even more damaging given that his overall leadership is a style that Hersey and Blanchard would define as a directing style—high directive and low supportive. Indications of this style and its negative aspects emerged in Portland when his style was described as "dictatorial," and by his contradictory and confusing response to cell-phone usage policy. However, the style was particularly highlighted in the high-profile tense and fear-ridden period of the sniper crisis, in which directive was the least desirable style to adopt. Two factors are characteristic of his approach: the way he handled data information during the sniper investigation and the way this information was shared (or not shared) with the involved law enforcement parties. Especially telling were the accusations of his

withholding a critical piece of information and (by doing so) endangering patrol officers who were searching for the suspects. Although his behavior was not malicious in nature but rather an error of judgment, such mistakes are highlighted by increased publicity and scrutiny. To some extent, the directive style is somewhat inevitable in the complex environment of this high-profile investigation, which involved numerous law enforcement bodies and competing jurisdictions. In order to be able to display a more confident style of leadership (e.g., S2 or S3), one must have insight into one's subordinates and shine less of a spotlight on their every move.

The desire to keep subordinates on a tight leash, minimizing empowerment, is frequently an outcome of a particular necessity rather than an ego trip. This directing approach can only be received well coming from somebody who is perceived as very knowledgeable and experienced in a situation in which subordinates truly lack the readiness to complete a task. While Moose had the experience, he demonstrated a lack of knowledge with regard to the steps to be taken in order to solve the case. Less task direction and more coaching and delegation of the subordinates would have resulted in much more support for him and his way of leadership during this time of crisis of national proportion.

Lastly, in the case of the sniper investigation, the display of weakness on national television only exacerbated the shortcomings of Moose's S1 approach. The public are not the only ones who suffer from such a breakdown; police officers, who look to their leader for guidance and coaching, rather than just dry directions on how to do things, are also likely to be alienated by the directive style.

After leaving Montgomery, Maryland, Charles A. Moose moved to Oahu, Hawaii and joined the Honolulu Police Department (HPD) as a rookie in 2006 (Former, 2006; Londono, 2006). He graduated from the HPD academy in November 2006 and began patrolling the streets of Honolulu as a rookie on November 13, 2006, and still works there today (Former, 2006; Londono, 2006).

CAREER HIGHLIGHTS
Situational Leadership Style (S4)

Chief Bernard Parks
Los Angeles Police Department (LAPD), California

Bernard Parks was born in Beaumont, Texas, on December 7, 1943 ("Full biography," 2003). He received a bachelors of science degree from Pepperdine University ("Full biography," 2003). Parks then earned a master's degree in public administration at the University of Southern California in 1976. He entered the LAPD in 1965 and is a veteran of the department's Internal Affairs Division (Wilborn, 2002).

Bernard Parks served as chief of the LAPD from 1997 to 2002. Upon becoming chief, Parks outlined several goals for the LAPD, including reorganizing the command staff, increasing the budget to support officer fitness programs, improving the working relationship between the department and other city agencies, and launching a type of Command Accountability Strategies (COMPSTAT) program in order to develop better crime prevention strategies (McGreevy, 1997). Regarding the latter, the chief indicated a desire to focus on "quality-of-life" crimes and gang activity in order to improve neighborhood security (Stewart, 1997). Parks also persuaded the police commission to designate money to improve police facilities in order to make stations more attractive for citizens. Improvements included access ramps for the disabled and correction of unsafe building conditions (Berbeo, 1998e). Reportedly, the former police chief, Willie Williams, had refused to develop any of these community-policing techniques (Stewart, 1997).

Regarding the reorganization of command staff, Parks eliminated the department's three assistant chief positions in order to have more direct control over the rank and file (Morgan, 1997). He also implemented a mandatory four-hour training session on ethics and workplace violence in an effort to reduce lawsuits against the police (Berbeo, 1998a). In addition, Parks established a Financial Crimes Division to focus on white-collar crimes such as major fraud, trademark infringement, computer crime, and elder abuse. He indicated that he was able to devote officers to these crimes because the violent crime rate was lower than in many years. Parks also initiated a recruitment drive to increase the force by 1,000 officers (there were 9,500 officers on the force in 1998) using federal COPS MORE II funds (Berbeo, 1998b).

In 1998, Parks successfully urged the police commission to add automatic handguns and rifles to the arsenal of patrol weapons available to specially trained officers. According to Parks, the weapons would be more accessible to officers during critical events. The idea stemmed from the February 28, 1997, North Hollywood bank robbery shootout in which officers were outgunned by robbers; 10 officers were wounded in the incident ("Police commission," 1998).

After his first year as chief, Parks received good reviews from community leaders. The human relations commission indicated that community members largely approved of Parks. A similar opinion was expressed by members of the Los Angeles Urban League. Parks attributed the good reviews to his revision of use-of-force methods, training, and policies. In the 18 months preceding August 1998, only three use-of-force complaints had been entered against police (Berbeo, 1998c). At the same time, Parks supported the Special Investigation section of the department, an elite detective squad that was trained to use extreme force and was heavily armed. The unit had been criticized for threatening the safety of bystanders during their operations and was the subject of several federal lawsuits. *The Los Angeles Times* reported that the squad had wounded dozens of innocent people during its previous 33 years of operation, yet the unit made an average of only 45 arrests per year ("Critics question," 1998). Parks also defended LAPD's pursuit policy from criticism following an October 14, 1997, crash during a police chase that killed a woman and seriously injured her daughter. He indicated that to outlaw police chases in the city would allow criminals to run amok. He also stated that it would be irrational to make a change in policy based on the emotion evoked by a particular crash (Koval, 1997).

In addition, Parks raised the standards of conduct for officers, according to an independent report by the Los Angeles Police Commission. The improvements included policies that help to identify at-risk officers and a range of disciplinary responses to address poor conduct (Berbeo, 1998d). Parks focused on discipline as a way of addressing the recommendations of the 1991 Christopher Commission (sparked by the beating of Rodney King). He started a pilot program to put video cameras in patrol cars to discourage misconduct and to clear officers who may be wrongly accused of abuse (McGreevy, 1998).

Due to the above-noted focus on officer discipline, Parks struggled to keep good relations with the rank and file ("Bernard Parks," 1997). Many negatively characterized Parks as a strict disciplinarian (Berbeo, 1998a). Conflict between the Los Angeles Police Protective League (LAPPL) and Parks over when and how officers should be punished plagued his tenure. The LAPPL claimed that officers were terrorized by the possibility that they would be disciplined or lose their job over minor infractions. The league backed reform that proposed an appeal process to a binding arbitration panel acting independently of Parks (Hutchinson, 1998). The morale problem was particularly evident in the results of a survey compiled by PriceWaterhouseCoopers for the city in December 2000. Officers from a variety of ranks indicated that they limited their response to radio calls because they felt that senior management would not support them if citizens made complaints. The year's statistics supported this phenomenon: In 2000, arrests dropped 25% from the previous year. Parks was singled out by many respondents as the cause of morale problems, and 32.5% of respondents said the best response to the problems would

be to oust senior management (Schodolski, 2000). A 2001 study by the University of California at Los Angeles and the University of Southern California showed that 57% of police officers would leave the department if they could. In January 2002, the LAPPL passed a vote of no confidence against Parks; according to the union, 93% of its membership did not support him. This vote led to demands for his resignation (Wilborn, 2002).

In 1999, one of the biggest scandals in Los Angeles history, Rampart, was exposed (Rogers, 1999). After pleading guilty to stealing cocaine from police evidence rooms in exchange for a lighter sentence, Rafael Perez revealed that officers at the Rampart Station of the LAPD routinely framed gang members for crimes, lied in court, and planted weapons on or shot unarmed people (Rogers, 1999). Perez said that officers awarded plaques to each other after wounding or killing people ("L.A.'s ungovernable," 2000). The scandal resulted in the city being sued for millions of dollars in civil liability lawsuits ("Key dates," 2002). Parks's early response consisted of suspending 10 officers and a supervisor at the Rampart Station (Rogers, 1999). The scandal, however, evolved to an investigation of 70 officers ("L.A.'s ungovernable," 2000). As a result, in over 100 cases, convictions were overturned or charges dropped. The scandal increased citizen mistrust ("Bratton calls for probe," 2003). Despite the scope of the Rampart affair, and the initiation of state and federal investigations, Parks defended his protocol for handling corruption ("L.A.'s ungovernable," 2000).

In 2002, Parks reluctantly signed a federal consent decree to address the problems exposed by the Rampart scandal ("LAPD complying," 2002). Specifically, the U.S. Department of Justice accused the LAPD of committing a pattern of civil rights violations that began before the Rampart scandal and of failing to initiate effective reforms ("Key dates," 2002). The required reforms consisted of improved officer training, a computer system to track officers' performance, new oversight of the antigang unit, and a ban on racial profiling during traffic stops. An independent monitor was hired who would have access to personnel documents until 2005 ("LAPD complying," 2002).

In February 2002, the year that Parks's five-year contract[1] was up, Mayor James K. Hahn announced that he was opposed to giving Parks a new contract. The mayor stated that he felt Parks was unable to address the insular culture of the department, racial profiling, recruitment, and community policing. According to Mark Ridley-Thomas, a black member of the city council, as well as civic and clergy leaders of the community, black constituents in general were angered by the announcement that Parks's contract should not be renewed (Almeida, 2002; Orlov, 2002a; Sterngold, 2002). The mayor's opinion also ran counter to reports issued by Chief Deputy City Attorney Terree Bowers and Councilwoman Cindy Miscikowski, who both indicated that the progress toward reform of the police initiated by the federal consent decree was on track ("LAPD complying," 2002). In response, Parks accused Hahn of lobbying the city council against him during contract talks (Orlov, 2002b). A *Los Angeles Times* poll in the spring of 2002 indicated that 50% of the residents supported the decision not to reappoint Parks. Hahn joined the LAPPL in claiming that Parks failed to address the problem of officer morale and was too resistant to necessary reforms. On April 22, 2002, Parks resigned, effective May 4, 2002, after the city council did not renew his contract. He could have stayed as chief until his term ended in August 2002 (Almeida, 2002).

In March 2003, Parks took office as a Los Angeles City Councilman for the 8th district. Among his priorities, he planned to address homelessness and housing problems (Villegas, 2003).

The overall motto and goal of Chief Parks appear to be the enhancement of the working conditions of police officers in an attempt to lower the crime rate and the improvement of the relationship between the police and the public. Those noble goals are no small vision for a department plagued by a history of officer-community tension as well as a real legacy of misconduct (given allegations made during the Rodney King and O.J. Simpson trials). In light of LAPD's past, it is rather naïve to assume that by replacing a number of high-ranking officers or by instituting some new modules of training that anyone would actually achieve a major change in the department and/or in the quality of relationships between the police and the public.

Furthermore, it is hard to achieve the desired level of integrity simply by instituting or introducing new mechanisms of control and new disciplinary sanctions. If one wants to maintain a department of integrity, first one must be aware of the way officers feel about mechanisms of control and sanctions before making them standard operating procedures. The Rampart scandal exemplifies, better than the overall dissatisfaction of officers and the unions, the low directive and low supportive style of Parks's leadership.

The major task during his tenure was the identification of priorities for the force and the identification of supervisors to enforce and supervise those priorities. This type of leadership would require the S2 type of situational leadership, coaching, in which a leader is highly involved in both the tasks and the socioemotional needs of subordinates. Instead, Parks identified many tasks without assigning clear priorities. It was unclear from his laundry list of changes what was the first problem to target—officers' benefits, police-community relations, decreasing crime rates, or police corruption. Although these are all interrelated, as the old adage goes, "You have to pick your battles." One cannot fight on all fronts at the same time; priorities have to be identified and then fully supported. In this case, success required less empowerment to senior command, a much higher degree of supportive involvement, and a higher directive approach to the problem of low morale and quite blatant misconduct. Parks's assessment of his command staff was erroneous and resulted in the choice of a wrong level of intervention. Instead of S2, a more laissez-faire type (S4) was chosen, for which his command staff was not ready. The consequences snowballed, resulting in the Rampart scandal and ultimately preventing the renewal of Parks's contract.

After leaving the LAPD, Parks unsuccessfully ran for mayor of Los Angeles in 2005 (Bernard

Parks, 2011). In 2008, Parks unsuccessfully ran for the L.A. County Board of Supervisors (Bernard Parks, 2011). He was later elected City Councilman (Bernard Parks, 2011). As City Councilman, he drafted an ordinance to govern fast food restaurants across Los Angeles (Bernard Parks, 2011). Parks's also drafted an ordinance to provide tax incentives to grocery stores if they provided their customers with fresh and healthy foods (Bernard Parks, 2011). He also established numerous youth programs in South Los Angeles including the Prevention Intervention and Education program at

---

[1]After the 1992 riots, the Christopher Commission recommended, and voters approved, a change to the city charter that took away civil service protection for the chief of police. Instead, chiefs are limited to a five-year term, renewable at the police commission's discretion (Wilborn, 2002).

Crenshaw High School (Bernard Parks, 2011). He is also currently involved in the Challengers Boys and Girls Club, the Los Angeles Urban League, and the Brotherhood Crusade (Bernard Parks, 2011).

Exercise

The police chiefs detailed in this chapter have been examined through the prism of situational theory. Using what you have learned from other theories in this book or from the summaries (as indicated in the Introduction) you have read thus far, it is possible to consider their approaches through the lens of different theories. What other theories can you find applicable to these chiefs? Are there lessons to be learned from those applications, or do those theories indicate alternative paths or approaches that might have yielded a more successful outcome?

References

Almeida, C. (2002, May 3). After bitter battle, police chief's last day in office Saturday. *The Associated Press*, state and local wire.

Barnett, J. (2002, October 10). Eyes turn to former Portland Chief Charles Moose, fills a difficult role in the search for Maryland sniper. *The Oregonian*, p. A1.

Bass, B.M. (1990). *Bass and Stogdill's handbook of leadership: A survey of theory and research*. New York: Free Press.

Berbeo, D. (1998a, July 24). LAPD officers get brush-up on ethics, workplace violence at training session. *Metropolitan News Enterprise*, p. 8.

_____ (1998b, July 30). Parks says LAPD will bolster its efforts against white collar crime. *Metropolitan News Enterprise*, p. 10.

_____ (1998c, August 20). Parks tells panel 80 percent of Christopher Commission reforms achieved. *Metropolitan News Enterprise*, p. 8.

_____ (1998d, August 26). Parks says reforms proposed in two audits have been implemented. *Metropolitan News Enterprise*, p. 10.

_____ (1998e, November 4). Police panel approves department's list of structural improvements needed. *Metropolitan News Enterprise*, p. 10.

Bernard Parks. (1997, August 26). *City News Service*.

Bernard Parks, Los Angeles city councilman. (2011, February 12). *KPCC*.Retrieved May 31, 2011, from http://www.scpr.org/news/2011/02/12/bernard-parks-member-city-council/.

Bernstein, M. (1999a, May 27). Police chief resigns. *The Oregonian*, p. A1.

_____ (1999b, June 29). Governor gives Moose a special day, praise for community policing work. *The Oregonian*, p. B2.

_____ (1999c, July 28). Moose closes out 6-year tenure of achievement, pitfalls as chief. *The Oregonian*, p. A1.

Blanchard, K. & B. Nelson (1996). Where do you fit in? *Incentive*, 170 (10), 65–66.

_____, D. Zigarmi & R. Nelson (1993). Situational leadership after 25 years: A retrospective. *Journal of Leadership Studies*, 1 (1), 22–36.

_____, P. Zigarmi & D. Zigarmi (1985). *Leadership and the one minute manager: Increasing effectiveness through situational leadership*. New York: William Morrow.

Bratton calls for independent Rampart probe. (2003, February 28). Retrieved June 22, 2003, from kcal9.com/topstories/topstoriesla_story05919851.html.

Bryman, A. (1986). *Leadership and organizations*. London: Routledge & Kegan Paul.

Chadbourne, J. (1980). Training groups: A basic life cycle model.*Personnel and Guidance Journal* , 59 (1), 55–58.

Charles Moose: Controversial police chief. (2003, October 10). BBC News UK Edition (online). Retrieved July 9, 2004, from http://news.bbc.co.uk/1/hi/world/americas/3129388.stm.

Clines, F.X. (2002, October 7). Fear in sniper's wake: "This guy's our neighbor." *The New York Times*, p. A12.

Community Policing Consortium. (1997). *Supervising the problem-solving process: A guide for supervision and supervisory instruction.*Facilitator's guide, version 2.0. Washington, D.C.: U.S. Department of Justice, Office of Community Oriented Policing Services.

Critics question whether police unit infamous for shoot-outs should be disbanded. (1998, November 30). *The Associated Press*, state and local wire.

Dainty, P. (1986). Leadership and leadership research. *Journal of Managerial Psychology*, 1 (1), 40–43.

Duin, S. (2002, October 27). He frayed, fumbled, but chief Moose rose to the challenge of D.C. sniper case. *The Oregonian*, p. A1.

Fermino, J. & M. Kranes (2002, October 23). Maryland's Moose a New York kid. *The New York Post*, p. 5.

Fernandez, C.F. & R.P. Vecchio (1997). Situational leadership theory revisited: A test of an across-jobs perspective. *Leadership Quarterly*, 8 (1), 67–84.

Former Maryland police chief joins Honolulu police. (2006, November 11). *KPUA.*Retrieved May 31, 2011, from http://www.kpua.net/news.php?id=9858.

Full biography for Bernard C. Parks. (2003, February 23). Retrieved from http://www.smartvoter.org/2003/03/04/ca/la/vote/parks_b/bio.html.

Graeff, C.L. (1983). The situational leadership theory: A critical view. *Academy of Management Review*, 8 (2), 285–291.

_____ (1997). Evolution of situational leadership theory: A critical review. *Leadership Quarterly,*8 (2), 153–170.

Greimel, H. (1998, November 19). Police to keep using beanbag ammunition in crowd control. *The Associated Press, state and regional.*

Hersey, P. (1985). *The Situational leader.* Escondido, CA: Center for Leadership Studies, Inc., in Community Policing Consortium (1997). *Supervising the problem-solving process: A guide for supervision and supervisory instruction.*Facilitator's guide, version 2.0. Washington, D.C.: U.S. Department of Justice, Office of Community Oriented Policing Services.

_____ (1977). *Management of organizational behavior: Utilizing human resources*(3rd ed.). Englewood Cliffs, NJ: Prentice Hall.

_____ & K.H. Blanchard (1969). Life cycle theory of leadership. *Training and Development Journal*, 23, 26–34.

_____ (1988). *Management of organizational behavior: Utilizing human resources* (5th ed.). Englewood Cliffs, NJ: Prentice Hall.

_____ (1993). *Management of organizational behavior: Utilizing human resources* (6th ed.). Englewood Cliffs, NJ: Prentice-Hall.

Holly, D. (1999, August 2). Moose vows to take on all issues to build better Montgomery County Police Department. *The Associated Press*, Washington dateline.

Hutchinson, E.O. (1998, October 18). Chief effort to limit police violence. *The Daily News of Los Angeles*, Viewpoint.

Kennedy, H. (2002, October 10). Angry over tarot leak police chief rips media, ex-cops for taunting killer. *Daily News*, p. 7.

_____ & M. Becker (2002, October 26). Moose is the toast of the beltway. *Daily News*, p. 6.

Key dates in LA police leadership. (2002, April 10). *The Associated Press*, state and local wire.

Korten, D.C. (1962). Situational determinants in leadership structure. *Journal of Conflict Resolution*, 6, 222–235.

Koval, A. (1997, October 16). Chief: If chases weren't allowed, we wouldn't want to be in this city. *City News Service*.

Kuykendall, J. & P.C. Unsinger (1982). The leadership styles of police managers. *Journal of Criminal Justice*, 10, 311–321.

L.A.'s ungovernable police. (2000, March 2). *The New York Times*, p. A26.

LAPD complying with federal consent decree reforms, city says. (2002, February 14). *The Associated Press*, state and local wire.

Londono, E. (2006, October 29). Montgomery's ex-chief to be a rookie in Honolulu. *Washington Post.*Retrieved May 31, 2011, from http://www.washingtonpost.com/wp-dyn/content /article /2006/10/28 /AR2006102800444.html.

Malkin, M. (2003, March 3). Charles Moose remains a celebrity months after sniper probe, but faces ethics review. *The Associated Press*, domestic news.

Manning, S. (2003, June 19). Maryland chief who led sniper probe quits. *The Day*. Retrieved January 13, 2011 from http://news.google.com/newspapers?id=Q55GAAAAIBAJ&sjid=l_gMAAAAIBAJ&pg=6209,3633395.

McGreevy, P. (1997, August 13). Parks takes office: New chief says Los Angeles is "moving forward." *The Daily News of Los Angeles*, News.

_____ (1998, September 26). Parks wants video cameras in all new police cars. *The Daily News of Los Angeles*, News.

Morgan, S. (1997, August 22). New Los Angeles police chief promises integrity. *Copley News Service*, state and regional.

Niehouse, O.I. (1984). Controlling burnout: A leadership guide for managers. *Business Horizons, 27* (4), 80–85.

Nkrumah, W. (1999, May 28). Neighbors show range of opinion on Moose. *The Oregonian,* p. D2.

Northouse, P.G. (2004). *Leadership: Theory and practice* (3rd ed.). Thousand Oaks, CA: Sage.

Orlov, R. (2002a, April 25). Mayor defends decision on Parks. *The Daily News of Los Angeles,* p. N3.

_____ (2002b, May 4). Parks out of LAPD at 5 p.m. *The Daily News of Los Angeles,* p. N4.

Oshiro, G.R. (1999, April 9). Portland-area chiefs denounce racist auto stops. *The Oregonian,* p. A1.

Parade honors officers who hunted sniper pair: Chief Moose in lead as grand marshal. (2002, November 24). *The Record,* p. A11.

Plan puts Portland police chief in charge of school police. (1999, January 20). *The Associated Press,* state and regional.

Police commission approves wider use of automatic weapons by officers. (1998). *Metropolitan News Service,* p. 11.

Reddin, W.J. (1967). The 3-D management style theory. *Training and Development Journal, 21* (4), 8–18.

Rogers, J. (1999, September 18). Police corruption scandal expands in Los Angeles: Hundreds of cases could be tainted. *The Record,* p. A18.

Schodolski, V.J. (2000, December 31). LAPD survey blames low morale on chief: Officers admit slacking off to stay out of trouble. *The Record,* p. A21.

Sisk, R. (2002, October 25). Finally it's hats off to the chief: Moose defiant, emotional. *Daily News,* p. 28.

Son of city's police chief is arrested on drug charges. (1997, September 15). *Seattle Post-Intelligencer,* p. B2.

Sperry, P. (2003, April 9). Police union probes Moose. *World Net Daily.* Retrieved May 16, 2003, from www.worldnetdaily.com/news.article.asp?ARTICLE_ID=31951.

Statement of Montgomery County executive Douglas M. Duncan regarding the resignation of police chief Charles Moose. (2003, June 18). Montgomery County Maryland Press Archives. Retrieved July 9, 2004, from http://www.montgomerycountymd.gov/apps/News/statements/ST_details.asp?StID=19.

Sterngold, J. (2002, February 6). Fraying ties with blacks, Los Angeles mayor opposes rehiring police chief. *The New York Times,* p. A14.

Stewart, J. (1997, August 21). News flash: He's an adult. After Willie Williams, Bernard Parks could come off as a genius simply by doing the obvious. *New Times Los Angeles,* Columns.

Stogdill, R.M. (1974). *Handbook of leadership: A survey of theory and research.* New York: Free Press.

Terzian, P. (2003, March 26). Commentary—Justice for Chief Moose. *Providence Journal-Bulletin,* p. B5.

Top sniper cop in tome-ly deal. (2003, January 23). *The New York Post,* p. 12.

Tsao, E. (2003, January 23). Police chief shares lessons of snaring D.C. area snipers. *The Sunday Oregonian,* p. B4.

Van Auken, P. (1992). Leading in four arenas. *Supervision, 53* (3), 9–12.

Video Monitoring Service. (1997, November 5). *KPTV-TV.* Retrieved May 18, 2003, from Lexis-Nexis.

Villegas, M. (2003, July 2). Ludlow takes oath as 10th district councilman: Four new members, plus Bernard Parks, are sworn in as city council gets new look. *Wave West,* pp. 1–8.

Vital stats: Chief Charles Moose. (n.d.). Retrieved May 16, 2003, from www.bet.com/articles/l,clgb4346_5042,00.html.

Washington, W. (2002, October 26). The sniper case arrests: Sniper probe earns praise and criticism. *The Boston Globe,* p. A1.

Wilborn, P. (2002, April 9). LA police commission votes against second term for Chief Parks. *The Associated Press,* state and local wire.

Yukl, G. (1994). *Leadership in organizations* (3rd ed.). Englewood Cliffs, NJ: Prentice Hall.

Doing Things Right or Doing the Right Thing
Contingency Theory

OVERVIEW OF THE THEORY

Contingency theory posits that certain leaders are compatible with certain situations, depending on variables such as leader–member relations, task structure, and position power. The theory is examined through the prism of two chiefs who were faced with extremely high-profile incidents: Chief O'Brien of the Miami Police Department (MPD), who had to manage the highly politicized Elian Gonzales case, and Chief Koby of the Boulder Police Department, who oversaw the murder investigation of Jonbenet Ramsey. The question for contingency theory is whether their particular mix of styles was right for managing these situations.

Fiedler's contingency theory is one of the most widely researched leadership theories, yet it is also controversial (Saha, 1979). Most studies indicate that the model is of limited or no success in predicting leadership effectiveness (Hill, 1969; Kuehl, 1977; Peters et al., 1985; Saha, 1979; Schriesheim et al., 1994; Utecht & Heier, 1976); however, it has some empirical support from meta-analytic studies (Strube & Garcia, 1981). A particular strength of this model is that it makes intuitive sense. "Careful observation of leadership behavior indicates that no one style of leadership behavior is always effective; sometime an authoritarian person is very successful, at other times he is a failure" (Hill, 1969, p. 35).

The theory represents an application of Lewin's (1951) postulation that behavior is a function of the interaction between the person and the situation (person × situation; Strube & Garcia, 1981). The primary concept of the theory is that task-oriented leaders will perform more effectively under very

---

Fiedler's Contingency Theory

- It is the most widely researched leadership theory.
- It has limited success in predicting leadership effectiveness.
- The theory makes intuitive sense.
- Careful observation indicates that no one style of leadership behavior is always effective; sometimes an authoritarian person is very successful, and at other times he is a failure.

Work Groups

- *Coacting groups* work independently on common tasks.
- *Counteracting groups* work together to negotiate and reconcile conflicting opinions and processes.
- *Interacting groups,* the focus of the theory, have situational favorableness based on three variables: leader–member relations, task structure, and position power.

(Fiedler, 1967)

favorable or very unfavorable conditions. Relationship-oriented leaders will be most effective in situations of moderate favorableness (Fiedler, 1967). According to Rice and Kastenbaum (1983), Fiedler defines situational favorableness as:

> The degree to which the situation enables the leader to exert . . . influence and control over his/her group situation. Treatment of the situation then became a matter of identifying specific variables that affect situational favorableness and formulating appropriate rules for combining such variables into a meaningful metric of situational favorableness. (pp. 377–378)

However, as Rice and Kastenbaum (1983) explain, the situational variables are measured via that which is assumed to affect situational favorableness: leader–member relations, task structure, and position power. Furthermore, the theory assumes that the interaction between the leader measured by the Least Preferred Coworker [LPC] scale) and the situation accounts for most of the variance in a group's performance (Kabanoff, 1981).

Fiedler assumes that work groups can be classified as coacting, counteracting, or interacting. Coacting groups work independently on a common task. Counteracting groups work together, but to negotiate and reconcile differences in viewpoint and approach (Fiedler, 1967). Interacting groups are the focus of the theory because these groups have situational favorableness based on the three variables of leader–member relations, task structure, and position power (Hill, 1969). To sum up the above, leaders behave in either a task- or relationship-oriented style, which is appropriate or inappropriate given the situational favorableness (based on leader–member relations, task structure, and position power).

Furthermore, in developing the Leader Match program, Fiedler (1998) sought to train managers to engineer a situation to match their particular leadership style. The training program assumes that leaders' styles are difficult to change; therefore, situations are what can be altered

Leader Orientation

Behavior is a function of the interaction between the person and the situation.

- *Task-oriented* leaders will perform more effectively under very favorable or very unfavorable conditions.
- *Relationship-oriented* leaders will be most effective in situations of moderate favorableness.

(Lewin, 1951)

Leader Match Program

- *Increase L-M relations.* Spend more time.
- *Decrease L-M relations.* Spend less time.
- *Increase task structure.* Break down tasks into smaller steps.
- *Decrease task structure.* Include more members in decision-making process and inform the higher-ups about the challenges.
- *Increase position power.* Acquire expertise and depend less on subordinates.
- *Decrease position power.* Give more responsibility to group members, share decision-making powers, and deemphasize organizational ranks.

(Fiedler, 1998)

(Schriesheim et al., 1994). Fiedler (1998) provides the following concrete suggestions for engineering a situation:

1. *To increase leader–member relations.* A leader can spend more time with subordinates, organize after-work activities, increase availability, listen attentively to subordinates' problems, share information, and reward subordinates.
2. *To decrease leader–member relations.* A leader can spend less time with subordinates, see subordinates only by appointment, keep contact business-oriented, avoid becoming involved in subordinates' personal problems, and bring new members into the group.
3. *To increase task structure.* Leaders can outline tasks and break them down into smaller steps and seek formal training.
4. *To decrease task structure.* Leaders can include group members in planning and decision making and can let higher-ups know the challenges and problems their unit faces.
5. *To increase position power.* Leaders can acquire expertise in order to depend less on subordinates, ask for higher-up approval on decisions that affect their unit, exercise power fully, make sure all information passes through them as leaders, and set a good example.
6. *To decrease position power.* Leaders can give more responsibility to group members, share decision-making powers, and deemphasize organizational ranks.

MILITARY LEADERS AND CONTINGENCY THEORY

Utecht and Heier (1976) studied whether the contingency model could predict successful military leadership. They tested the main component of contingency theory that task-oriented leaders will perform more effectively under very favorable or very unfavorable conditions, while relationship-oriented leaders will be most effective in situations of moderate favorableness. Military success for the purposes of this study was defined as an officer who has attended or is attending a top military academy. Data were collected by questionnaire from classes of Air Force, Army, and Navy war colleges. Part of the questionnaire consisted of the LPC in order to elicit information about leadership style. Questions about the respondent's last six leadership functions measured task structure information. Rank status inferred position power. Chi-square analysis of the data yielded results indicating that the model was no better than chance in predicting military success. From this, the authors postulate that the model does not work in high structure settings such as the military.

They further speculate that this conclusion could be generalizable to "structured non-military organizations" such as the police (Utecht & Heier, 1976, p. 618).

COGNITIVE RESOURCES THEORY

Cognitive resources theory was developed by Fiedler (Fiedler & Garcia, 1987), and posits that intelligent, competent leaders make more effective plans, decisions, and strategies, which leaders communicate through directive behaviors. Initially, Fiedler indicated that intelligence was irrelevant to the effectiveness of leadership; a potential weakness of the theory. However, later on, he added the variables of intellectual ability, competence, and experience to the contingency model (Vecchio, 1990). However, Fiedler continues to draw a distinction between possessing intellectual capacity and being able to effectively harness it (Fiedler, 1996). Cognitive resources theory is an improvement on pure trait theory, which views traits in a vacuum. With cognitive resources theory, situational forces are important because they can have differential effects on the trait–performance association. The theory implies that leaders should be selected based on their valued traits of expertise, experience, and intelligence, and that work situations should be engineered to have low stress and group support of the leader (Bryman, 1986).

LEADER SUBSTITUTES THEORY

The leader substitutes theory is based on the notion that group processes can produce effects similar to interpersonal hierarchical leadership (Tosi & Kiker, 1997). The idea was developed by Kerr and Jermier (1978) from the observation that situational factors can actually negate a leader's ability to be effective. The model makes a distinction between two kinds of situational variables: substitutes (what makes leader behavior unnecessary or redundant) and neutralizers (what nullifies the effects of the leaders; Yukl, 1994). For example, in a highly structured organization, any attempt on the part of the leader to increase structure is neutralized by the existing structure.

Kerr and Jermier (1978) even suggest that leaders may be superfluous in some situations. They seem to mean that formal leadership may be unnecessary in the presence of capable subordinates.

VROOM–YETTON CONTINGENCY MODEL

Vroom and Yetton's (1973) contingency model focuses on the autocracy-participation dimension of leadership style. It aims to determine the impact of participative leadership on the quality of decision making. The model looks at leadership as a problem to be solved. In order to handle the problem, the leader may or may not share his/her decision making with subordinates. The level of participation represents a continuum of five levels of employee participation, from none to democratic participation. The leader decides which of the five levels to adopt based on seven rules, which are designed to either protect the quality of the decision or protect the acceptance of the decision by subordinates. According to the model, a single leader should be flexible enough to vary his/her style across the continuum as needed; therefore, a person has the capacity to be both autocratic and participative (Bryman, 1986). Contrastingly, Fiedler's contingency theory assumes that leadership style is fixed.

DYADIC APPROACH TO CONTINGENCY THEORY

Vecchio (1979) integrates Fiedler's contingency model with Graen's notion of vertical dyadic linkage. Here, the focus is on the interaction that occurs between each group member and the leader. Leaders are "discriminating among and dealing with them as individuals, such that certain subordinates receive leader-exchange opportunities while others are merely supervised" (Vecchio, 1979, p. 591). Subordinates fall into one of two categories: the "trusted assistants" (the in-group) and the "ordinary members" (the out-group). Under this theory, low LPC leaders will be more productive relative to high LPC leaders when the environment contains both good and poor leader–member relations. Moreover, leaders are differentially affected by the groups and alter their styles depending on whether they are interacting with an employee in the in- or out-group. Out-group members are best motivated by a leader who emphasizes extrinsic motivators (external pressures on the group which demand performance) while in-group members perform better when motivated by intrinsic factors (e.g., personal commitment to the leader).

CRITICISMS OF CONTINGENCY THEORY

To begin, the theory assumes that factors that affect situational favorableness are congruent with each other in a given situation. There is no prescription for how a leader should respond when one part of a situation suggests one style but another part of the situation suggests the opposite (Bryman, 1986). The second criticism is the problem of ambiguous causal time order because empirical research into contingency theory tends to look at static relationships under the assumption that leader style affects leader performance when it could be the other way around, that leader performance affects leader style.

CONTINGENCY THEORY OF LEADERSHIP AND THE CASES OF CHIEF WILLIAM O'BRIEN AND CHIEF TOM KOBY

The contingency theory of leadership, in essence, states that not all leaders are good for all situations. The theory attempts to match the best possible leader to a given situation. Situations in police environments can be complex and extreme, for example, because of political pressure (such as that faced by Chief William O'Brien during the Elian Gonzales problem) or sensitivity due to the extremely tragic nature of a case (such as that faced by Chief Tom Koby during the Jonbenet Ramsey situation). In these types of cases, only a specific individual possessing certain qualities and skills can truly be a leader. All those with a different orientation are destined to fail, primarily due to the nature of the situation.

Leadership during a particular situation can be measured in two ways. It can be seen as a failure from an organizational perspective (e.g., the police organization within the pyramid of local politics). The same result can also be seen as a success for an individual within this system (e.g., on a personal moral level). Whether such a moral victory makes a person a bad leader depends on which perspective is more important—organizational or individual. In this book about police leadership, the focus is on police organizations, not on individual police officers. The examples focus on individuals as they reflect the nature of police work generally rather than the work of one specific police officer. As a result, it is sometimes the case that a police chief whom this author holds in high esteem has nevertheless for those very reasons failed according to the contingency theory and the organizational view of leadership.

As in the situational and style theories, contingency theory researchers focus on the interaction between the leader and the situation, and they delineate between task- and relationship-oriented behaviors. What is different with contingency theory, and particularly with the Vroom–Yetton contingency model (1973), is that the phenomenon of leadership is considered a problem to be solved. Of course this depiction of leadership can be and probably is applicable to every chief described in this book; however, both Chief O'Brien and Chief Koby faced very specific events that presented them with unique problems to be solved. Both had other difficult cases in their careers that they managed differently. These particularly high-profile incidents and the interpretation of them were chosen specifically to illustrate this theory.

Police corruption and misconduct, riots, increases in crime, or even serial killers are the bread-and-butter problems faced by police leaders on a daily basis. However, the purely political and internationally covered case of a child who has to be returned to a parent in another country or the high-profile murder of a child of very wealthy and influential parents who themselves become the major suspects—these cases are indeed the problems that defined the success or failure of those chiefs' leadership. No other problems either chief faced in his career as a police leader were as critical. Chief O'Brien and Chief Koby are prime illustrations of the Vroom–Yetton contingency model's definition of leadership as a function of the success or failure with which a problem is solved.

The only argument one might have with this author is about the distinction between success and failure. Although Chief O'Brien was the man for the Gonzales situation and he handled it properly, he was simply not the right man for the larger existing political configuration. He put the idea of policing before the idea of police organization. Although on an individual and moral level he was a successful leader, based on the contingency model and in the sense of responsibility first to the organization, he failed. Also according to contingency theory, Chief Koby failed in handling the Jonbenet Ramsey case; he was simply not the right man for the situation.

CAREER HIGHLIGHTS

Contingency Theory

Chief William O'Brien and the Elian Gonzales Case
Miami Police Department, Florida

William O'Brien was born in 1944 and raised in LaGrange, Illinois. He earned his bachelor's degree in political science from the University of Miami (Driscoll, 2000). From 1968 to 1973, he was an Air Force pilot, attaining the rank of captain and served in the Vietnam War ("Biography," 2000; Driscoll, 2000). From 1975 to 1998, he was an officer in the MPD, spending 18 years on the Special Weapons and Tactical (SWAT) team ("Biography," 2000). O'Brien started the first sexual battery unit in the department and also served as commander of Internal Affairs for more than three years (Driscoll, 2000).

O'Brien served as chief of the MPD from 1998 to 2000. On taking the job in 1998, O'Brien cautioned his officers to "maintain their integrity and stay away from politics" (Driscoll, 2000, p. 1B). Officers told the press that under his leadership the department's morale was high. They also indicated that the equipment improved and that training was more comprehensive. O'Brien was described as "a professional leader" by one sergeant (Driscoll, 2000). However, others criticized O'Brien for being merely a puppet to the city manager, Donald Warshaw, who previously held the post of police chief (Epstein, 1998).

O'Brien appeared open to outside review of the department. O'Brien supported one federal investigation of five police officers accused of planting a gun on an innocent man ("Federal investigation," 1998). He also supported the city's investigation of the records of a crime-prevention group, Do the Right Thing, which was run by Warshaw when he was chief ("City," 1999). In April 2000, O'Brien suspended two officers without pay; one was suspected of causing the death of an inmate, and the other was a sergeant who ordered that the report be falsified to make the death seem accidental ("Police officer," 2000).

The war on drugs preoccupied the MPD during O'Brien's tenure, and the department often worked with federal agencies. In January 1999, 26 federal indictments were handed down for drug dealing in connection with a gang known as the John Doe Boys. O'Brien was quoted as saying he hoped that such a massive drug bust would act as a general deterrent (Cherfils, 1999).

In January 1999, the chief joined with black religious leaders to plead with a drug gang member to surrender after perpetrating 11 shooting deaths in the previous five months (Kidwell, 1999). Indeed, early in his tenure, O'Brien reached out to the African-Caribbean community in Little Haiti, a particularly crime-prone neighborhood. O'Brien pledged to increase police presence (Casimir, 1998). However, O'Brien also took criticism from the African-American community, particularly after the shooting of an unarmed 19-year-old African American by an officer in September 1999. The officer reportedly shot the victim because he believed the victim was reaching for a gun. African-American activists called for a federal investigation, although O'Brien indicated he believed the shooting was justified (Zarella, 1999).

O'Brien's tenure as police chief was overshadowed—and ultimately doomed—by the Elian Gonzales incident that divided Miami in 1999 and 2000. Elian, a six-year-old Cuban child, was a shipwreck survivor rescued off the coast of Florida on November 25, 1999. The ship intended to secretly dock in Florida to facilitate illegal Cuban immigration. Elian's mother perished in the immigration attempt. His Cuban-American relatives in Miami hoped to keep the boy in the United States (as his deceased mother clearly had wished to immigrate), at the home of relative Lazaro Gonzales, stating that Elian had a right to live in freedom in the United States. However, his father in Cuba sought his return, and as a Cuban citizen, Elian was potentially deportable. During the federal appeals process, the Cuban community in Miami staged several protests in support of Elian's relatives ("Elian," 2000). The federal government, under the direction of then Attorney General Janet Reno, staged a raid on April 22, 2000, to seize Elian from his Miami relatives (Davis & Black, 2000). Tens of thousands of demonstrators took to the streets in Miami to protest and counterprotest in the weeks after the seizure ("Elian," 2000). More than 300 people were arrested, and tear gas was used for crowd control ("Miami in turmoil," 2000).

According to press reports, Cuban-American citizens initially criticized O'Brien over the police response to a January 6, 2000, exile protest during which the police used tear gas (Jimenez, 2000). The main issue, however, stemmed from the seizure of Elian on April 22, 2000. Federal authorities alerted O'Brien to the raid before it happened. Yet O'Brien informed neither Mayor Joe Carollo nor City Manager Don Warshaw. Furthermore, one of O'Brien's top assistants, who had been the police liaison to the federal government during the investigation, accompanied federal agents on the raid, even though the mayor had pledged that the Miami police would not help in the federal plan to seize Elian (Branch & Bridges, 2000). Reportedly, O'Brien was reluctant to tip off Carollo because Carollo had openly sided with Elian's relatives and O'Brien was worried that telling him would compromise the mission (Bragg, 2000a). After the raid, O'Brien stated, "This was a police issue, not a political issue" ("Mayor sacks manager," 2000). He also said, "If word had gotten out, there would have been a confrontation astronomically greater than there was, putting not only law enforcement personnel at risk but the people in [Elian's] house at risk, the child himself, the demonstrators, and even the media at the scene" (Bragg, 2000a, p. A9). However, Carollo, a Cuban American, was infuriated by O'Brien's failure to disclose information about the raid and told the city manager to fire O'Brien. Warshaw refused to fire O'Brien and so Carollo fired Warshaw. However, Carollo stated the firing had nothing to do with Elian's case but rather with Warshaw's pattern of lying to him ("Mayor sacks manager," 2000).[1]

On April 29, 2000, O'Brien resigned, saying he could no longer work for a divisive mayor. The resignation was emotional, with O'Brien shedding tears ("Miami in turmoil," 2000). Cuban exiles applauded the mayor's desire to fire O'Brien (Cubans make up 60% of the city's voters). People from other groups (Caucasians, African Americans, and some Hispanics), however, expressed criticism, saying Carollo was not observing the rights of anyone but Cuban Americans (Bragg, 2000a). In his resignation speech, O'Brien took responsibility for the events associated with the seizure of Elian (Jimenez, 2000). When O'Brien resigned, officers showed their support for him. The INS officer who led the Elian raid called O'Brien a "hero of the mission" by maintaining operational security.

The new police chief, Raul Martinez, became the first Hispanic to hold the position in Miami. Warshaw, who was appealing his termination by Carollo and who was in the 10-day grace period for terminating the position, hired Martinez despite Carollo's desire to conduct a nationwide search for a chief. Carollo then accused Warshaw of misusing public funds, blackmailing him, and "grabbing at anything he can to keep himself in power" (Bragg, 2000b). By May 10, 2000, Carollo had replaced City Manager Warshaw with Carlos Gimenez, who was previously the city's fire chief. This added to the Cuban-American majority on the city commission (now three out of five; Bragg, 2000b).

When Chief O'Brien warned his officers to stay away from politics when he took command of the police department, it was almost a premonition of what was to come in his career. For generations, police chiefs and champions of the idea of policing,

both academics and practitioners, have expressed the desire to separate policing from politics and politics from policing. No matter how noble this concept and how much one believes in its importance for creating an effective, efficient, and, above all, truly professional police force, it is absolutely unachievable. Police always were and will forever remain a tool in the hands of the ruling government. Ideally, should the police be separated from government as far as professional decision making is concerned? Yes, they absolutely should be. Unfortunately, these strictly professional decisions will probably remain only in the sphere of the acquisition of new equipment and the choice of color and shape of new uniforms. Everything else will remain political, to either a greater or a lesser degree, maybe, but still political.

A friend recently confronted this author; he said the word "political" has become a disparaging term because anything and everybody can be portrayed as doing something political. This criticism is valid. There can be no doubt that everyone is motivated by some overt or covert agenda, consciously or subconsciously. Therefore, it becomes imperative to clarify what exactly is meant by the word "political" in any given context where it is used. In the context of Chief O'Brien and in the context of policing in general, when the word "political" is used, it means that the police chief or commissioner is asked to engage in action or inaction by the local politician in charge. In Miami, it was the refusal of Chief O'Brien to inform the mayor, who happened to be sympathetic to the community backing one side in the controversy, about the decision to act. O'Brien's decision was based on a strictly professional assessment of the situation at hand (the problem) and devoid of any political considerations. His professional assessment made him the right man for the problem, but his lack of political consideration made him the wrong leader for the case. From the individual standpoint, he made the right personal choice; looking at the police as a profession, he made the right professional choice. But from the organizational standpoint—looking at police organization as the tool of the local government and therefore, by default, being accountable to it—he failed.

Following his tenure as Chief of the MPD, O'Brien was named acting chief of the DeKalb County Police Department in February 2009 (Matteucci, 2010). He was later appointed Police Chief in Dekalb County on October 13, 2010 (Matteucci, 2010). O'Brien is currently in the midst of dealing with a budget crisis and has been ordered to cut $4.7 million dollars from the department's budget (Matteucci, 2011). To meet this requirement, O'Brien is considering doing some or all of the following: implementing furlough days for police officers; grounding the department's police helicopter; and eliminating the 80 recruits who would graduate from the police academy in the summer of 2011 (Matteucci, 2011).

[1] According to municipal laws, only the city manager can fire the police chief.

CAREER HIGHLIGHTS

Contingency Theory

Chief Tom Koby and the Jonbenet Ramsey Case
Boulder Police Department, Colorado

Tom Koby was born in 1949 in Cleveland, Texas. He grew up in Houston, joining its police department in 1969. In 1985, he was the commander of the burglary and theft division and earned the Manager of the Year award. By 1987, he was deputy chief and, four months later, assistant chief. Lee P. Brown, the chief who promoted Koby, described him as an "intelligent, bright and energetic police officer. He cared for the police officers under him. He understood that you have to be out in the community" (McPhee, 1997a, p. B4). While working as a police officer, Koby earned his bachelor's degree in business administration in 1973 (McPhee, 1997a).

In May 1991, Koby became chief of the Boulder Police Department, a force with 213 employees (140 officers); he served until 1998. The town of Boulder had 100,000 residents at the time he took over the force. Koby was reportedly hired because of his commitment to community-oriented policing (McPhee, 1997a). One change he implemented was to assign one officer per shift to specifically handle calls and complaints about burglar alarms in order to cope with the 8% of calls that were false alarms. This freed up other officers to form additional patrols of downtown and the campus of Colorado University (McCullen, 1996). In addition, because of his concern for order maintenance, Koby opposed the opening of a methadone clinic in downtown Boulder, stating that it would bring drug problems into the city (McCullen, 1997f). He had observed increased crime, such as drug peddling and the selling of stolen goods, near Houston methadone clinics. However, in the first few weeks of the clinic's opening, no problems were reported (Robey, 1998).

Koby tagged underage drinking as one of the larger concerns in Boulder, a college town. He believed that many crime problems in the community could be attributed to alcohol. Rather than giving out more driving under the influence (DUI) tickets, Koby reached out to community leaders (including sorority and fraternity members) at Colorado University to encourage a change in campus attitudes about drinking (McPhee, 1997a). However, on May 2, 1997, a riot erupted on campus. During the riot, students started fires and threw bottles. Police officers responded with tear gas and rubber bullets. Several people were injured and 35 people were arrested (McPhee, 1997b). Approximately 1,500 people had gathered on campus during the unrest ("Hundreds fight," 1997). Reportedly, the melee was fueled by student frustration with the city and school administrations over efforts to control underage drinking. Beer was no longer permitted in the stadium during football games. Fraternity and sorority leaders claimed they were coerced into banning liquor in 1995. But the most commonly cited reason for the unrest was that a fraternity had its charter revoked because of alcohol-related violations (McPhee, 1997b). Some sources reported that the riot also stemmed from a celebration of the end of the semester ("Hundreds fight," 1997). Additional unrest occurred two nights later when a crowd leaving a theater began lighting fires and breaking windows. Police controlled them with tear gas and rubber bullets ("Hundreds fight," 1997).

Regarding the riots, Koby told the press, "Friday night was a mob mentality. Blaming it on students not being able to drink as much as they wanted is crap. This is about drunks. They got drunk and tore the town up to show they have the right to get drunk" (McPhee, 1997b, p. B1). Koby subsequently met with campus leaders to encourage them to help stop the violence by handing out flyers and chalking the sidewalks with messages to remain calm (McPhee, 1997b).

Even as early as 1994, Koby faced police union criticism over his lack of communication with his executive staff, lack of enough staffing in the communications center, and lack of training in the department. In May 1994, Officer Beth Haynes was killed in a shootout. A dispatcher did not relay information that there was an armed man involved in the unfolding incident, and backup was slow to assist her; the police union held Koby responsible (McCullen, 1994).

Koby's tenure as police chief was consumed by the investigation into the death of 6-year-old child beauty pageant star Jonbenet Ramsey, who was strangled in her home on December 26, 1996. That morning, police responded to a call by her mother, Patsy Ramsey, that Jonbenet had been kidnapped; Jonbenet's mother allegedly found a ransom note. Later that day, Jonbenet's father, John Ramsey, and a neighbor found the body of the dead girl in the basement of the Ramsey's home (McPhee & Chronis, 1997). In the early stages of the investigation, Koby provided few details to the public, indicating his frustration with the national media attention the case received[1] (Richardson, 1997). For example, by late January 1997, he would only say that there were more than seven suspects, but little else (Brooke, 1997a). Koby also expressed confidence in the ability of his officers to handle the investigation (Richardson, 1997). His decisions in the case were publicly supported by the city council and city manager. Unlike in most homicide cases, Koby debriefed the city manager regularly on developments (McCullen, 1997b).

Koby was criticized for the police handling of the case in the early stages of the investigation. Specifically, questions arose as to why John Ramsey and a neighbor were permitted to look for Jonbenet when this activity is usually limited to the police. Unsupervised by authorities, John Ramsey picked up his daughter's body when he found her and moved her upstairs, thus disturbing the crime scene (Richardson, 1997). Moreover, the Ramseys and five friends had free rein throughout the home in the seven hours before Jonbenet's body was found, thus threatening the integrity of evidence at the scene. For example, the autopsy results showed that Jonbenet's skull was fractured by a blunt instrument. The first patrol officers at the house noticed a heavy flashlight on the kitchen counter, but it was missing by the end of the day (Brooke, 1997b). Despite this, on January 9, 1997, Koby denied that the moving of the body destroyed evidence at the crime scene (Richardson, 1997). He stated, "I've been in communication with police personnel around the country and most legal experts will tell you we've done it quite right" (McCullen, 1997a, p. A4).

Further criticisms involved the special treatment allegedly given to the Ramseys. Detective Linda Arndt, the first detective on the scene, was accused of "bonding" with the parents rather than keeping them under suspicion, as is standard practice when conducting such an investigation (McPhee, 1997a). In addition, the Ramseys were not formally interviewed immediately after the crime, which drew national criticism (McPhee, 1997a). Additionally, Arndt placed a sheet over the body of Jonbenet as it was being taken out of the house, potentially disturbing evidence (Brooke, 1997b). The department was also criticized for the delay of two and a half hours between Patsy Ramsey's 911 call to police and the arrival of a detective (though patrol officers responded immediately; Brennan & McCullen, 1997). In mid-May 1997, Arndt was taken off the investigation, reportedly because fewer detectives were needed. More likely, it was her inappropriate compassion for the Ramseys that made Koby lose confidence in her. This lack of skepticism about the Ramseys was cited as the reason it took seven hours to find Jonbenet's body (Brennan & McCullen, 1997).[2]

Koby also refused help from the Denver Police Department, which investigates approximately 100 murders a year, compared to the 1 or 2 that occur annually in Boulder[3] (Brennan & McCullen, 1997). This inexperience led to poor decisions; for example, the police were going to turn the home back over to the Ramseys just hours after the body was discovered—not a usual procedure in homicide investigations. Moreover, Koby refused an FBI offer to monitor the Ramseys, their relatives, and their associates for any suspicious activity. In fact, prior to the Ramsey investigation, Koby had never led a murder inquiry (Brennan & McCullen, 1997).

In one of the most controversial investigative decisions, Koby was criticized for using Jonbenet's body as a bargaining chip to secure an interview with her parents. The Boulder Police told the parents they would only get her body back for burial if they gave formal statements. Then the police miscalculated by agreeing to submit written questions to the Ramseys instead of holding a normal police interrogation during which police have the additional tactical edge of using follow-up questions and psychological techniques (Brennan & McCullen, 1997).

Internal and external conflict also characterized the Ramsey investigation. The police and the district attorney's office did not get along; in one instance, the police withheld DNA evidence from the district attorney (Brennan & McCullen, 1997). The reasons were not disclosed to the press but were interpreted as reflecting a conflict (Brooke, 1997b). The evidence was finally handed over six weeks after the police received it (Richardson, 1997). However, on September 15, 1997, Koby and District Attorney Alex Hunter met with private attorneys advising the team on the case in an apparent show of cooperation (Brennan, 1997b). In addition, detectives on the case had conflicting attitudes and strategies (Brennan & McCullen, 1997). For example, there were allegations that leaks to the media were coming from Sergeant Larry Mason;[4] however, this may have been a retaliatory accusation related to the different strategy that Mason advocated. On the first day of the investigation, he was vocally critical of other detectives for not separating the Ramsey parents, not interviewing them in depth, and not ridding the home of milling friends and family in order to conduct a proper search (Brennan & McCullen, 1997).

In May 1997, Boulder Police who were union members delivered a no-confidence vote on Koby over a variety of issues. Subsequently the city manager, who was also criticized for supporting Koby, resigned (Richardson, 1997). The primary issues were his response to the riots at Colorado University (McCullen, 1997c) and his implementation of community-oriented policing ("Chief," 1998). Regarding community-oriented policing, officers complained about the department moving away from basic police services to an ineffective "feel good" department (McCullen, 1997e). Critics also said Koby mismanaged the media in the Ramsey case ("Chief," 1998), although there was no specific reference to the Ramsey investigation in the documentation of the no-confidence vote (McCullen, 1997c).

More than 10 months after the Ramsey investigation began, in October 1997, Koby admitted that the early stages of the investigation were not handled well, in contradiction to previous comments. "If we had to do it all over again, we would do it differently," Koby told reporters

("Belated acknowledgement," 1997, p. A41). At that time, Koby replaced the commander in charge of the investigation, James Eller, with Mark Beckner (who would later succeed Koby as chief; McCullen, 1998c). Koby also added three more detectives. Yet the press in a *Rocky Mountain News* editorial believed that the reorganization would do little to save the mishandled case ("Belated acknowledgement," 1997). Beckner did help to promote the investigation team as hardworking (McCullen, 1997g). An editorial in *The Denver Post* similarly opined that the mishandled case could not be salvaged (Green, 1999). Koby also lamented that the case dominated police time and that only one-quarter of the 1,150 criminal cases requiring investigation received any attention in 1997 (Brooke, 1998).

On November 19, 1997, Koby announced his plan to retire in December 1998. The date was set to permit him enough time to wrap up the Ramsey investigation (Robey, 1997; "The Jonbenet Ramsey case," 2001). The police union president indicated, however, that within the ranks, anti-Koby and pro-Koby factions were developing that were threatening cohesiveness in the department. By setting a resignation date, the union president believed Koby wanted to mitigate these divisions (McCullen, 1997d). On March 31, 1998, however, Koby was asked to continue as police chief and to act as a special project coordinator for the city manager. Mayor Bob Greenlee indicated that Koby's announcement that he would resign in December 1998 resulted in an even further decrease in confidence because the department appeared unstable. "When you set a course, people need to be confident you will keep steering the course," Greenlee said (McCullen, 1998b, p. A5). Koby worked on special projects for the city manager until approximately December 1998, when Koby went into retirement ("Former Boulder cop," 1999). A year later, Koby left retirement to take a position at Tango Partners, a Boulder firm that invests private resources in Internet technology companies and then gives the proceeds to nonprofit charities ("Former Boulder cop," 1999).

Despite the departure of Koby, the Ramsey investigation continued to languish. Hundreds of thousands of dollars were spent during the 16-month investigation, producing 30,000 pages of documents. A grand jury failed to indict anyone for Jonbenet's murder by the end of its term in October 1999 ("Former Boulder cop," 1999). The investigation by the police department finally ended in February 2003 when the case shifted to being solely the domain of the district attorney's office (Weller, 2003).

One of the most frightful situations in police work deals with parents who have lost their child to a violent crime. The only thing that is more frightful is to consider those parents as prime suspects. If the parents also happen to be very powerful socialites from a relatively small local community, nobody can envy the job of the local police chief. Nonetheless, if Chief Koby had engaged in task-oriented behavior instead of relationship-oriented behavior prior to the murder case, his leadership style could have saved him from major criticism. After his appointment as chief of police of the Boulder Police Department, he had introduced many initiatives with the officers' and the community's well-being in mind. However, this relationship orientation took him too far away from the task of the police organization. The task was not just to maintain a good relationship with the community and introduce community-oriented programs but also to deliver professional policing, based inter alia on well-trained and experienced police officers. Koby closed Boulder's police academy and sent his officers to regional academies. This measure alone took away from his ability to have any input into the ways officers are and should be trained. This move was motivated by cost savings for the community, but it was not a task-oriented move and later prevented Koby from being able to handle the Jonbenet Ramsey investigation in a more professional manner.

During the investigation, when his officers did not handle the crime scene in a proper manner, Koby had his relationship orientation put to the test. When he further refused help from the FBI, his task orientation became even more diluted.

His lack of effective cooperation with the district attorney's office might have earned him some support from his subordinates but certainly did not contribute to the task at hand. The murder case was never solved, and even if his relationship with the community had not been strained, it may not have saved him from criticism. Both his mismanagement of the murder case and the "feel good" approach to community policing contributed to the demand for his resignation. The biggest problem he faced—the murder case of Jonbenet Ramsey—required a leader whose primary orientation was task rather than relationship. Koby's relationship orientation cost him his career as chief of the Boulder Police Department.

Exercise

The police chiefs detailed in this chapter have been examined through the prism of contingency theory. Using what you have learned from other theories in this book or from the summaries (as indicated in the Introduction) you have read thus far, it is possible to consider their approaches through the lens of different theories. What other theories can you find applicable to these chiefs? Are there lessons to be learned from those applications, or do those theories indicate alternative paths or approaches that might have yielded a more successful outcome?

[1] Since resigning as police chief in 1998, Koby has not made any public statement concerning the murder case (Brennan et al. 1999).

[2] Arndt later sued Koby, the subsequent police chief Mark Beckner, and the city for ruining her reputation as an investigator by taking her off the case ("McCullen", 1998e). She also complained that Koby failed to defend her against criticisms, using her as a scapegoat for a failing investigation (McCullen, 1998a, 1998d; "Trial begins," 2001). By not allowing her to speak to defend herself, Arndt allegedly had her First Amendment rights violated. In 2001, after the case went to trial, the judge ruled that she failed to prove that she was prevented from speaking to defend herself ("Ex-detective's suit," 2001; "Trial begins," 2001).

[3] Of the 16 homicides that occurred between 1990 and 1997, 11 were solved by the Boulder Police Department, a percentage only slightly higher than the national average (McCullen, 1998d).

[4] In September 1997, Koby intended to submit the Ramsey detectives and their superiors to lie detector tests to find out who might have been leaking information to the media; however, the contract with the police union prohibited polygraph tests of officers (Brennan, 1997a).

References

Belated acknowledgement. (1997). *Rocky Mountain News*, p. A41.

Biography of Chief William O'Brien. (2000, April 28). *The Associated Press*, state and local wire.

Bragg, R. (2000a, April 29). Miami police chief quits in raid fallout. *The New York Times*, p. A9.

_____ (2000b, May 10). Legacy of a Cuban boy: Miami city hall is remade. *The New York Times*, p. A20.

Branch, K. & T. Bridges (2000, April 23). Carollo slams chief over role. *The Miami Herald*. Retrieved June 15, 2003, from www.miami.com/mld/miami/news2059832.htm.

Brennan, C. (1997a, September 20). Koby cancels lie detector tests. *Rocky Mountain News*. Retrieved August 21, 2003, from http://denver.rockymoutainnews.com/extra/ramsey/0920jon/htm.

_____ (1997b, September 16). Boulder officials hold talks: DA, police meet over Jonbenet murder. *Rocky Mountain News*, p. A4.

_____ & K. McCullen (1997, June 8). Ramsey case a tragedy of errors: Inexperience, sympathy, unwillingness to ask for help combine to trip up investigation. *Rocky Mountain News*, p. A4.

_____, McCullen, K. & L.G. Everitt (1999). Koby looks like "someone who has retreated from the world." *Rocky Mountain News*, p. A21.

Brooke, J. (1997a, January 31). After slaying of girl, family and town are in uneasy limbo. *The New York Times*, p. A17.

_____ (1997b, July 2). Case of Jonbenet Ramsey stalls on error and rivalry. *The New York Times*, p. A14.

_____ (1998, April 22). Grand jury refocuses spotlight on Jonbenet case. *The New York Times*, p. A12.

Bryman, A. (1986). *Leadership and organizations*. London: Routledge & Kegan Paul.

Casimir, L. (1998, December 17). Chief on the beat: Miami's O'Brien reaches out. *The Miami Herald*, p. 8.

Cherfils, M. (1999, January 7). U.S. attorney's office unseals 26 indictments in gang wars. *The Associated Press*, state and local wire.

Chief who led initial investigation into Ramsey murder gone. (1998, December 31). *The Associated Press*, state and local wire.

City asked for review of anti-crime group's records. (1999, September 19). *The Associated Press*, state and local wire.

Davis, P. & C. Black (2000, April 30). 1,000 march in Miami protest over Elian. *CNN.com*. Retrieved June 15, 2003, from www.cnn.com/2000/US/04/29/cuba.boy.03.

Driscoll, A. (2000, April 29). Bill O'Brien bids farewell. *The Miami Herald*, p. 1B.

Elian protests split Miami. (2000, April 29). *BBC News Online*. Retrieved June 12, 2003, from http://news.bbc.co.uk/2/hi/americas/730885.stm.

Epstein, G. (1998, October 22). Miami's new chief: Veteran of streets called "right man for the job." *The Miami Herald*, p. 1B.

Ex-detective's suit dismissed. (2001, June 13). *The New York Times*, p. A31.

Federal investigation targets Miami cops. (1998, December 14). *The Associated Press*, state and local wire.

Fiedler, F.E. (1967). *A theory of leadership effectiveness*. New York: McGraw-Hill.

_____ (1996). Research of leadership selection and training: One view of the future. *Administrative Science Quarterly*, 41 (2), 241–250.

_____ (1998). The leadership situation: A missing factor in selecting and training managers. *Human Resources Management Review*, 8 (4), 335–350.

_____ & J.E. Garcia (1987). *New approaches to leadership: Cognitive resources and organizational performance*. New York: John Wiley.

Former Boulder cop to work for philanthropic firm. (1999, December 29). *The Associated Press*, state and local wire.

Green, C. (1999, October 14). So now a killer walks. *The Denver Post*, p. A1.

Hill, W. (1969). The validation and extension of Fiedler's theory of leadership effectiveness. *Academy of Management Journal*, 12, 33–47.

Hundreds fight police near Colorado campus. (1997, May 5). *The New York Times*, p. A12.

Jimenez, J.L. (2000, May 4). Elian made a tough guy cry. *Miami New Times*, News.

Kabanoff, B. (1981). A critique of leader match and its implications for leadership research. *Personnel Psychology*, 34, 749–764.

Kerr, S. & J. Jermier (1978). Substitutes for leadership: Their meaning and measurement. *Organizational Behavior and Human Performance*, 22, 374–403.

Kidwell, D. (1999, January 1). Shooting suspect urged to surrender. *The Miami Herald*, p. B1.

Kuehl, C.R. (1977). Leader effectiveness in committee-like groups. *Journal of Business*, 50 (2), 223–230.

Lewin, K. (1951). *Field theory in social science*. New York: Harper.

Matteucci, M. (2010, October 13). Dekalb native to be named police chief. *The Atlanta Journal-Constitution*. Retrieved May 31, 2011, from http://www.ajc.com/news/dekalb/dekalb-native-to-be-681412.html.

_____ (2011, March 1). DeKalb votes against tax increase, cuts $33 million. *The Atlanta Journal-Constitution*. Retrieved May 31, 2011, from http://www.ajc.com/news/dekalb/dekalb-votes-against-tax-848354.html.

Mayor sacks manager who wouldn't fire chief. (2000, April 28). *The Associated Press*, state and local wire.

McCullen, K. (1994, June 15). Cops' union at odds with chief. *Rocky Mountain News*, p. A20.

_____ (1996, December 30). 15 new cops on the block in Boulder, 7 jobs added: More manpower due downtown, on the Hill. *Rocky Mountain News*, p. A22.

_____ (1997a, January 10). Chief defends investigation: Boulder police don't think serial killer is loose in city, deny preferential treatment. *Rocky Mountain News*, p. A4.

_____ (1997b, March 6). Ramsey investigation called "on track": Boulder city manager confident about inquiry, police chief says probe hasn't busted budget. *Rocky Mountain News*, p. A30.

_____ (1997c, May 23). Cops vote on confidence in chief, Boulder police react to management of riot involving CU students. *Rocky Mountain News*, p. A52.

_____ (1997d, November 19). Boulder police chief to retire in December 1998: Koby plans to stay until decision is made in Jonbenet slaying. *Rocky Mountain News*, p. A5.

_____ (1997e, September 22). Police union watches, waits on Koby: Chief says he's working to solve staffing, morale problems in department. *Rocky Mountain News*, p. A26.

_____ (1997f, October 30). Plan for methadone clinic disturbs Boulder officials: Police chief concerned with "bringing people with drug problems into the heart of the city." *Rocky Mountain News*, p. A35.

_____ (1997g, December 6). A different style of doing things: New lead investigator in Jonbenet case has a more open attitude in dealing with public. *Rocky Mountain News*, p. A5.

_____ (1998a, February 5). Detective taken off Ramsey case to sue: Investigator wants $150,000, claims chief won't defend her in media or let her speak out. *Rocky Mountain News*, p. A5.

_____ (1998b, April 1). Boulder police chief loses job: Acting city manager decides to move Koby to another post to work on special projects. *Rocky Mountain News,* p. A5.

_____ (1998c, September 21). Ramsey case pulls careers into "vortex": Inquiry's complex swirl has battered many people whose scars have yet to heal. *Rocky Mountain News,* p. A5.

_____ (1998d, January 18). Boulder's unsolved murders: Jonbenet's killing one in five in '90s being studied by police. *Rocky Mountain News,* p. A39.

_____ (1998e, September 24). Lawsuit against former Boulder police chief out on hold: Ramsey investigation prompts 90-day delay in action on Koby case. *Rocky Mountain News,* p. A38.

McPhee, M. (1997a, February 9). Koby weathers storm over his silence: Boulder chief known as manager, thinker. *The Denver Post,* p. B4.

_____ (1997b, May 6). Riot police keep uneasy peace in Boulder. *The Denver Post,* p. B1.

_____ & Chronis, P. (1997, January 1). Neighbors, police tight-lipped. *The Denver Post,* p. A1.

Miami in turmoil as police chief quits: Resignation came after mayor fires. (2000, April 29). *The Florida Times-Union,* p. A6.

Peters, L.H., D.H. Hartke & J.T. Pohlmann (1985). Fiedler's contingency theory of leadership: An application of the meta-analysis procedures of Schmidt and Hunter. *Psychological Bulletin,* 92 (2), 274–285.

Police officer, sergeant investigated in inmate's death. (2000, April 20). *The Associated Press,* state and local wire.

Rice, R.W. & D.R. Kastenbaum (1983). The contingency model of leadership: Some current issues. *Basic and Applied Social Psychology,* 4 (4), 373–392.

Richardson, V. (1997, July 14). Jonbenet's killer may never be found: Cops still hopeful, others lose faith. *The Washington Times,* p. A1.

Robey, R. (1997, November 19). Koby reveals retirement goal: Boulder chief sets date at end of '98. *The Denver Post,* p. B1.

_____ (1998). New clinic no trouble so far. *The Denver Post,* p. B2.

Saha, S.K. (1979). Contingency theories of leadership: A study. *Human Relations,* 32 (4), 313–322.

Schriesheim, C.A., B.J. Tepper & L.A. Tetrault (1994). Least preferred co-worker score, situational control, and leader effectiveness: A meta-analysis of contingency model performance prediction. *Journal of Applied Psychology,* 79 (4), 561–573.

Strube, M.J. & J.E. Garcia (1981). A meta-analytic investigation of Fiedler's contingency model of leadership effectiveness. *Psychological Bulletin,* 90, 307–321.

The Jonbenet Ramsey case day-by-day. (2001). Retrieved August 21, 2003, from http://us.cnn.com/US/9703/ramsey.case/timeline5.html.

Tosi, H.L. & S. Kiker (1997). Commentary on "substitutes for leadership." *Leadership Quarterly,* 8 (2), 109–113.

Trial begins in Jonbenet Ramsey case. (2001, May 30). *The New York Times,* p. A18.

Utecht, R.E. & W.D. Heier (1976). The contingency model and successful military leadership. *Academy of Management Journal,* 19 (4), 606–618.

Vecchio, R.P. (1979). A dyadic interpretation of the contingency model of leadership effectiveness. *Academy of Management Journal,* 22 (3), 590–600.

_____ (1990). Theoretical and empirical examination of cognitive resources theory. *Journal of Applied Psychology,* 75 (2), 141–147.

Vroom, V.H. & P.W. Yetton (1973). *Leadership and decision-making.* Pittsburgh, PA: University of Pittsburgh Press.

Weller, R. (2003, February 10). Boulder avoids suit with new probe into case. *The Associated Press,* state and local wire.

Yukl, G. (1994). *Leadership in organizations* (3rd ed.). Englewood Cliffs, NJ: Prentice Hall.

Zarella, J. (1999, September 27). Federal probe demanded after Miami police shooting. *CNN.com.* Retrieved June 16, 2003, from www.cnn.com/US/9909/27/police.shooting.

Winning Minds and Hearts
Path-Goal Theory

OVERVIEW OF THE THEORY

Path-goal theory matches leadership style with the characteristics of subordinates and the work environment. In this chapter, the theory is examined through the approach and innovations of Commissioner William Bratton, whose career included three major U.S. police forces: Boston, New York, and Los Angeles. It also looks at Chief Richard Pennington of the New Orleans Police Department (NOPD), who was charged with turning around a highly corrupt force.

Path-goal theory takes into consideration a combination of situational variables that combine to moderate the effects of leader behavior (Jermier, 1996). The leader chooses from four styles: directive, supportive, achievement-oriented, and participative (Northouse, 2004).[1] These leader behaviors are moderated by the personal characteristics of subordinates (e.g., preference for external structure, need for achievement, and self-perceived ability) and environmental factors (e.g., task structure, work group norms, and organizational formalization). These variables determine the best type of leader behavior to use, so subordinates can accomplish their goals. In essence, path-goal theory seeks to predict how situational variables will affect the impact of leader behavior on subordinates (Indvik, 1986).

Path-goal theory is derived from the expectancy theory of motivation articulated by Vroom (1964) and Georgopoulos, Mahoney, & Jones (1957). The expectancy theory states that subordinates' motivation is a function of their expectation that their behavior will lead to a specific outcome. Uncertainty and ambiguity, therefore, are stressful and counterproductive. Uncertainty makes it hard for subordinates to identify outcomes as they relate to personal utility. It also obscures the likelihood that outcomes can be predicted based on what the subordinates do to accomplish tasks (Dessler, 1972). In addition, another articulation of the theory proposes that subordinates choose levels of effort that they are willing to

---

[1] However, Wofford and Liska (1993) note that most of the empirical tests of the theory have focused merely on supportive versus directive styles. These styles are measured using the Leadership Behavior and Descriptor Questionnaire (LBDQ) family of scales, which were developed for style theory testing of consideration and structure. Therefore, achievement-oriented and participative styles have been largely ignored by researchers. Furthermore, Schriesheim and Von Glinow (1977) state that the dimensions of the Ohio Leadership Study scales differ greatly from the theory. They contain "punitive, autocratic and production oriented items … which are extraneous to the measurement of the theory's leadership constructs" (p. 399).

employ. A high level of effort only results from subordinates' perception that it leads to good performance (Bryman, 1986). In another formulation of the theory, subordinates adjust their level of effort or motivation in light of whether it will lead to personally desirable outcomes (higher pay, recognition of achievement, promotion, enjoyment) versus undesirable ones (layoff, stress, rejection by coworkers, reprimand, accident; Yukl, 1994).

Based on expectancy theory, House (1971) developed the path-goal framework, which combined leader structure, task uncertainty, and subordinate attitudes. Here, the leader attempts to clarify the goals of subordinates as well as the route subordinates should take to accomplish them.

Yukl (1994) provides a schematic of the causal relationships of directive and supportive leadership on subordinate effort. Causal variances (leader behaviors) plus intervening variables (subordinate expectancies) as well as situational variables (such as the characteristics of the task and environment and the characteristics of subordinates) lead to the end result of subordinate effort and satisfaction (Yukl, 1994). In directive leadership, if the leader reduces role ambiguity and increases expectations of effort and performance, or increases the magnitude of incentives and thus the desire for task success, or strengthens the reward contingencies and thus the performance reward, it will result in increased subordinate effort (Yukl, 1994). In supportive leadership, if the leader reduces boredom, it will increase the intrinsic valence of the work, or if the leader increases the workers' self-confidence and lowers anxiety, it will increase expectations regarding effort and performance and again lead to increased subordinate effort (Yukl, 1994).

Since the theory focuses on the needs of followers, the assumption is that leaders should modify their behavior accordingly. It focuses on followers' perceptions of whether a particular activity helps or hurts their likelihood of attaining a goal. In essence, subordinates' perceptions drive the leadership style and the resulting effectiveness of the organization.

The theory places a great deal of emphasis on the responsibility of leaders to adapt to subordinate circumstances on a day-to-day basis. "It reduces the emphasis on measurement of leadership styles and replaces it with an emphasis on the leader's breadth of available behaviors and sensitivity to the combination of situational variables" (Wofford & Liska, 1993, p. 875). Unlike in previous theories, in path-goal theory, the leader must be flexible, depending on subordinate demands in a given situation. Much like the situation in Fiedler's contingency theory, the leader must be able to make a situational diagnosis, which in the case of the path-goal theory is based on subordinate needs. The leader must 1) determine the key factors that would affect subordinate motivation, 2) identify the degree to which these factors are provided by the organizational environment, and 3) predict the effects of different types of leadership that could be employed. The theory also implies that there are situations in which leadership is irrelevant. Although House (1971) did not develop this corollary, Jermier (1996) states: "[A] major feature of the path-goal theory of leadership that made it interesting was that it bordered on denying the importance of leader behaviors in certain situations, thereby raising to awareness the fundamental assumption of the field" (p. 313). The fundamental assumption referred to here is that some form of leader behavior always impacts effectiveness.

Critics of the theory emphasize that the theory is very broad in explanations and encompasses too many possibilities. The reformulated theory developed by House (1996) includes ten classes of leader behavior, individual differences of subordinates, and task moderator variables. However, the meta-proposition of the theory does reduce to one simple concept.

> The essence of the theory is the meta-proposition that leaders, to be effective, engage in behaviors that complement subordinates' environments and abilities in a manner

that compensates for deficiencies and is instrumental to subordinate satisfaction and individual and work unit performance (House, 1996, p. 47).

PATH-GOAL THEORY AND MILITARY LEADERSHIP

Mathieu (1990) studied the moderating influence of subordinates' needs for achievement and affiliation as it relates to the relationship between instrumental and supportive leadership and subordinates' satisfaction. His sample consisted of 298 Army and Navy ROTC cadets (81% men and 19% women) who completed questionnaires asking them to rate their leaders. Results indicated that subordinates with a high need for achievement prefer leaders who clarify the path toward goals "such that their efforts may be best translated into performance" (House & Dessler 1974, p. 185). However, contrary to prediction, individuals with a high need for affiliation were more satisfied with instrumental leadership behaviors (as opposed to supportive leader behaviors). Mathieu (1990) states that this result could indicate that the task-related activities of structured leaders also provide opportunities for social interactions.

PATH-GOAL THEORY AND THE CASES OF COMMISSIONER WILLIAM BRATTON AND CHIEF RICHARD PENNINGTON

The path-goal theory aims to understand and predict how situational variables will affect leader behavior toward subordinates and focuses on how leaders can motivate subordinates to accomplish their goals. The leaders who adopt the path-goal approach tend to engage in one of the following styles: directive, supportive, achievement-oriented, or participative. The two police chiefs chosen to illustrate this leadership approach both chose different styles of the path-goal theory. Of course, their careers were far too complex to be generalized and summarized into a single category, but overall, for the purposes of illustrating the path-goal theory, Commissioner William Bratton of the New York Police Department (NYPD) adopted the directive leadership style, while Chief Richard Pennington of the New Orleans Police Department (NOPD) engaged in the achievement-oriented style.

The directive style sets standards of performance and rules for tasks, and it clarifies goals and expectations. According to this approach, the leader tells the subordinates what to do, when to do it, and how to do it. This more difficult part of the directive style can and should be sweetened by guidance and adequate supervision of subordinates. Such guidance and supervision should motivate the employees/police officers and should also transmit knowledge to those in need of such directives. The achievement-oriented style establishes a high standard of excellence whereby the employees are expected to excel based on the situational variables created by the leader.

Both styles require what Richards (2002) calls "the art of winning minds and hearts." Whether choosing the directive or achievement-oriented approach, one has to win both the minds and hearts of the subordinates in order to accomplish the clearly defined tasks (see Table 9–1). To win minds, the objective is to convince people and to make them understand the purpose of what they are asked to do, no matter how detailed and directive the explanation is. To make them excel, one has to win their hearts, where the objective is to move people to increase their commitment and motivation. The medium for winning minds is information; the medium for winning hearts is emotion.

Both Bratton and Pennington engaged in the art of winning minds and hearts despite the fact that both chiefs had most, but not all, of the leadership competencies needed to win both

TABLE 9-1   Art of Winning Minds and Hearts Leadership Competencies Needed

Art of Winning Minds	Art of Winning Hearts
Objective: Convincing people—make them understand the purpose	*Objective:* Moving people—increase their commitment/motivation
Medium: Information	*Medium:* Emotion
• Relevance	• Authenticity
• Cogency	• Empathy
• Showmanship	• Listening
• Questioning	• Rapport
• Dialogue	• Optimism
• Promotion of discovery	• Trust in intuition
• Atmosphere of intellectual safety	• Atmosphere of emotional safety

(Richards, 2002)

minds and hearts. They were not missing innate qualities or skills; the nature of police organization itself stymied some of those competencies. It is hard, if not impossible, to create intellectual and emotional safety in the punitive and semi-military structure of a police department. It is even harder to do in a police environment that is fraught with problems such as corruption, lack of accountability, poor training, and high crime rates. These challenges are topped by a civil service personnel structure in which hiring and firing cannot easily be tailored to the chief's vision. Things can be demanded, created, and implemented, but success will be limited without winning minds and hearts.

CAREER HIGHLIGHTS
Path-Goal Theory, Directive Leadership Style

Commissioner William Bratton
New York Police Department (NYPD), New York

William Bratton, whose career has included heading police departments in several major American cities, is likely best known for the sweeping changes he introduced while commissioner of the NYPD from 1994 to 1996.

Bratton was raised in the Dorchester section of Boston. He earned a bachelor's degree in law enforcement from Boston State College/University of Massachusetts. He is also a graduate of the FBI National Executive Institute ("William J. Bratton," n.d.) and has honorary doctorates from John Jay College of Criminal Justice, New York Institute of Technology, and Curry College (Dussault, 1999).

Bratton joined the Boston Police Department (BPD; Metropolitan District Commission Police) in 1970, where he rose through the ranks to become superintendent of police, the highest sworn rank, by 1980 ("William J. Bratton," n.d.). As a patrol officer in Boston, he was awarded the Schroeder Brother's Medal of Honor for talking down a bank robber and rescuing a hostage (Dussault, 1999). In 1983, he was chief of the Massachusetts Bay Transportation Authority and was credited with facilitating a 37% reduction in crime in the jurisdiction. Bratton returned to the Boston police in 1986, resuming his superintendent role.

He emphasized affirmative action initiatives and hired the first female chief of patrol. In 1990, he moved to his first position in New York City as the chief of police/senior vice president of the New York City Transit Authority. During his service there, subway crime was reportedly reduced by 50% ("William J. Bratton," n.d.) due to his quality-of-life policing campaign, which involved arrests of young New Yorkers for fare evasion as a method of reducing the occurrence of more serious crimes (Greene, 1999). As the *Economist* reported, as a harbinger of things to come,

> Mr. Bratton reinvigorated New York's transit police by improving equipment, but also through a dose of Druckerish "Management by objectives": the transit cop's task became to cut crime, not merely to respond to it. Morale soared, and subway crime was cut in half. ("NYPD, Inc.," 1995)

The following year, Bratton again returned to the BPD as superintendent in chief, serving from 1992 to 1993 (Newfield & Jacobson, 2000; "William J. Bratton," n.d.). He brought community-oriented policing to Boston and was supported by community leaders for respecting the concerns of community residents as this style of policing was introduced ("A test," 1993; Lakshmanan, 1993). He also added more police officers and computers to compile crime statistics by neighborhood. Bratton focused on reducing community members' fear of particular crime-prone blocks or neighborhoods in the city. Each officer received 40 hours of professional development in community-oriented policing in the first few months of Bratton's tenure. Recruits were required to attend a weeklong problem-solving training course ("New Boston police chief," 1993).

Boston officers generally supported Bratton, although his career path led to criticism and speculation that he wanted to return to New York City as police commissioner. He was said to be so ambitious that he was "willing to hurt a department, even a friend, to help himself" (McGrory, 1993,

p. 15). However, the executive director of the Police Executive Forum in 1993 stated that Bratton recognizes "people's strengths and weaknesses and delegates power and gives them tools. . . . He is a big picture person, who has five or six things going on at the same time. He doesn't micromanage. He understands policing" (McGrory, 1993, p. 15). Bratton did return to New York in 1994 to lead the NYPD.

In New York, Bratton is associated with the zero-tolerance style of policing, which emphasizes quality-of-life issues for the average law-abiding citizen by enforcing laws against petty drug dealers, prostitutes, graffiti artists, and "squeegee men." Bratton directed patrol officers to stop citizens and search them for violating minor laws, to conduct warrant checks, or to question citizens about criminality in their neighborhood. Bratton thus put into practice Wilson and Kelling's "broken windows" theory, which suggests that serious crackdowns on low-level disorderly behavior would reduce serious crime because such order–maintenance problems attract predatory criminals (Greene, 1999; Wilson & Kelling, 2011). Criticizing a 1995 cost-saving initiative city judges put forward for decriminalization of minor offenses, Bratton explained:

> Go out in Brooklyn any night of the week. All these characters out there on the corner, drinking, urinating. Once they get a few drinks under their belt, out come the knives, out come the guns, out come the loud mouths . . . little things count . . . little things like quality of life. One of the reasons why this city and this country were in [a] mess . . . was nobody paid attention to the small things." (Marzulli, 1995, p. 2)

Under Bratton, the number of misdemeanor arrests jumped from approximately 133,000 to 205,000 annually between 1993 and 1996 (Walker & Katz, 2002).

Commentators criticized Bratton for using military campaign metaphors ("Don't mention," 2003; Greene, 1999). In fact, Bratton described one anti-drug crime initiative, Juggernaut (never implemented), in the following terms:

> We would literally roll through the city, like securing a beachhead. And after we secured the beachhead, unlike the previous initiatives in this city, we wouldn't just let the area sink back; we would move in the occupation forces which would be the community police and the beat cops. (Newfield & Jacobson, 2000, p. 52)

Citizens, often minorities, complained about overzealous policing and brutality. The proportion of "general patrol incidents" made up 29% of the complaints in the last year of the Dinkins administration and increased to 58% the next year under Mayor Rudolph Giuliani and Bratton (Greene, 1999). Bratton reportedly responded to this increase by pointing out that the board dismisses 75% of the complaints as unsubstantiated prior to investigation. The increase in complaints also correlated with the creation of an independent (civilian), rather than internal, complaint board (McQuillan, 1995). However, the number of complaints to this Civilian Complaint Review Board increased more than 60% between 1992 and 1996. Regardless, relations with minority communities were reportedly particularly strained as a result of the zero-tolerance policing style. At a community meeting in the South Bronx in July 1995, a critical crowd of approximately 500 Hispanic people met Bratton and Giuliani, complaining about police brutality. Bratton told them, "You're acting like a bunch of fools. You're making a fool of yourselves. You're the same group that shows up at every one of these meetings and we're not getting anywhere" (Lewis & Siemaszko, 1995, p. 4). Yet in an earlier *New York Times* editorial regarding police brutality, Bratton got good marks for his response to the death in custody of Ernest Sayon. That victim had died after a struggle with the police, who restrained him using a controversial choke hold (banned by the NYPD). Bratton initiated a task force review of the department's policies and training programs on handling violent arrestees ("Mr. Bratton's wise policing," 1994).

Bratton strengthened command and discipline, and he emphasized officer accountability ("Community policing," 1994), making an analogy that he ran the police department as if managing a private business in which the profit is crime reduction. He devolved power to precinct commanders, whom he then held accountable for the crime statistics in their precincts. Initiating the computerized statistics (COMPSTAT) program, Bratton collected data on arrests, calls for service, and complaints, and he analyzed and mapped them at the precinct level. At weekly meetings, crime trends were examined and precinct commanders were questioned about their crime-control strategies. Those who could not reduce crime in their precincts were reassigned to less challenging positions (Walker & Katz, 2002). Bratton described these often-intense meetings:

> The reality is that COMPSTAT does produce some tension. If you go to school without doing your homework, you are scared to death the teacher is going to call on you. In COMPSTAT, you are guaranteed to be called on. So if you don't prepare, if you don't participate in the team, you are going to be embarrassed right out of the room. (Dussault, 1999)

As a result, there was high turnover in precinct commander positions. For the year prior to September 1995, three out of five precincts had new commanding officers (George, 1995).

Bratton also attacked corruption (Blood, 1995). Internal Affairs sting operations increased 10× from 50 to 500 investigations, over the year ending August 1995 (Lewis & Siemaszko, 1995). In addition, the department focused on the problem of cop alcoholism, a problem highlighted after a deputy inspector of a Brooklyn precinct showed up drunk at 7 a.m. for a strategy session with departmental bosses (George, 1995; Lewis & Siemaszko, 1995). That officer lost his command. That same year, 30 NYPD officers at a convention in Washington, D.C., went on a beer-soaked spree, groping women and running around naked. Bratton responded by devoting Internal Affairs resources to combating the cultural factors in policing that encourage both alcoholism and corruption; he also sent a series of self-help videotapes to the home of each officer in the city (Lewis & Siemaszko, 1995).

Late in Bratton's tenure, he publicly feuded with Mayor Giuliani (Wilborn, 2002). Bratton felt that Giuliani was taking all the credits for the much-lauded policing initiatives in order to bolster his own political ambitions (Newfield & Jacobson, 2000). He also said that the rift was caused by a difference in police strategies, which developed over the course of their working together. In a 1997 feature in *The New Yorker,* Giuliani said he had to do the work of the police commissioner because Bratton was too concerned with his media image to police effectively (Wilborn, 2002).

After leaving the police commissioner job in 1996, Bratton founded The Bratton Group, LLC, and was a consultant with Kroll Associates, which oversaw the federal consent decree with the Los Angeles Police Department (LAPD). Simultaneously, from 1993 to 1996, he was elected president of the Police Executive Research Forum (PERF; "William J. Bratton," n.d.). In October 2002, Bratton took the helm of the LAPD, which was reeling after the breaking of the Rampart scandal.[1] Many have speculated that Bratton was selected to run the LAPD because its image was so tarnished by corruption and he was known as a results-oriented police leader (Wilborn, 2002). Secondly, he had experience with the issues facing the city after serving on the committee overseeing the implementation of the federal consent decree. Finally, commentators have pointed out that Mayor James Hahn wanted to implement community policing as operationalized by Bratton in New York (Laird, 2002).

As expected, Bratton has decentralized authority to geographic bureaus within the department and updated the computer system to replicate New York's COMPSTAT (Stockstill, 2002). Bratton believes that increasing officers in problem areas such as South Central Los Angeles, where violent crime has dramatically increased, is the key to effective policing, but he has been granted only modest increases in force strength due to budget constraints in 2002 and 2003 (Broder, 2003; Nash, 2002).[2]

Bratton has made combating gangs foremost on the department's agenda, likening the gangs to the Mafia in New York. There are more than 200 gangs with 100,000 members operating in the Los Angeles (LA) area. The number of homicides in LA was the highest in the nation in 2002 (Leovy, 2003).[3] Most commentators attribute the homicide spike to increased gang activity, particularly by the Rolling 60s Crips (Kandel, 2002). Bratton promoted a captain as an anti-gang "czar" and has been encouraging officers to stop suspecting gang members for minor infractions, again using the "broken windows" theory to influence a reduction in serious gang activity. In November 2002, Bratton personally responded to two homicides, saying, "I want to go to homicide scenes so that I can be angry. I want to have intimacy with that," (Nash, 2002, N3).

When Bratton joined the NYPD, he was faced with a number of critical problems. The NYPD was a decentralized bureaucracy with very serious communication and accountability problems. Six months of logs containing reports of arrests, youth violence, and drug and gun activities all pointed to dysfunction and a lack of control, direct outcomes of overall bad management and an overwhelming lack of accountability. At the time, Jack Maple, one of the engineers of COMPSTAT, told this author that the commanding officers of the precincts had reached such a level of comfort in their inertia that when shown a map of their own precinct, they were not even able to identify it.

Bratton's approach to these serious problems was clearly one of a directive leader. He improved communication by dividing the organization into smaller entities to facilitate communication patterns and accountability at a precinct level; he instituted new programs to deal with youth violence and drug and gun activities; he fired many mid-level and senior managers; and he made it clear that promotions would be tied directly and exclusively to merit. The greatest innovation of all, with regard to personal and organizational accountability, was the now-infamous COMPSTAT, the statistical database that made precinct commanders and other ranking officers almost personally accountable for crime rates. However, despite the obvious advantages of COMPSTAT, it did not consider the human factor, the need for creation of both intellectual and emotional safety—minds and hearts—in order for police officers to buy into the concept. Bratton's idea of reengineering a police department was definitely a worthy effort. However, his path-goal approach, exhibited through the directive leadership style, did not factor in the missing component: One must buy the minds and hearts of one's subordinates when introducing highly innovative and extreme approaches. Attrition and elevated levels of cynicism were only some of the negative side effects of Bratton's directive style of leadership.

In 2006, LAPD Chief William Bratton had Officer Sean Joseph Meade arrested for assaulting a juvenile in a holding cell; the incident was caught on videotape ("LA Officer," 2006; "LA Police," 2006). One year later, Bratton had to deal with the fallout of the MacArthur Park incident ("Los Angeles Police Chief William Bratton," 2007; Prengaman, 2007a; Steinhauer & Preston, 2007). In response to that incident, Bratton criticized his officers handling of the clash with immigration protestors ("Los Angeles Police Chief William Bratton," 2007). According to Bratton, the protestors were throwing rocks and bottles at the LAPD officers who responded by striking protestors with batons and firing rubber bullets into the crowd, despite the presence of children (inappropriate use of force according to Bratton; "Los Angeles Police Chief William Bratton," 2007; Prengaman, 2007a). Numerous protestors and over a dozen police officers were injured during the confrontation; ten of whom had to be taken to the hospital for treatment ("Los Angeles Police Chief William Bratton," 2007; Steinhauer & Preston, 2007).

In the aftermath of the MacArthur Park incident, Bratton launched an investigation into the officer's conduct during the incident ("Los Angeles Police Chief Promises," 2007; Prengaman, 2007b). The Civilian Police Commission also opened an investigation ("Los Angeles Police Chief Promises,"

2007). Several days later, the FBI launched an investigation into the MacArthur Park incident alleging that the officers violated the civil rights of protestors ("FBI," 2007; "L.A. Mayor," 2007). The FBI investigation was supported by Bratton ("FBI," 2007). At the conclusion of the investigation into the MacArthur Park incident, Bratton demoted Deputy Chief Cayler "Lee" Carter Jr. and reassigned Carters second in command Louis Grey as punishment for how they handled their subordinates at MacArthur Park (Blankstein, 2007; Geis, 2007; Helfand, McGreevy, & Blankstein, 2007).

In 2008, Bratton opposed a task force created to protect famous individuals from bothersome tabloid photographers ("Celebs," 2008; "LA Police," 2008; "LA To," 2008; "Stars," 2008; Watkins, 2008).

On June 4, 2009, Bratton was awarded the honorary title of Commander of the Most Excellent Order of the British Empire by Queen Elizabeth II (Rubin, 2009a).

Later that year, he also opposed a proposal to halt the hiring of new police officers and to suspend recruitment of police officers until January 2010 (Rubin & Zahniser, 2009). According to Bratton, this move, although alleviating the city's budget deficit, countered his goal of increasing the size of the LAPD (Rubin & Zahniser, 2009).

Bratton resigned from the LAPD on October 31, 2009 (Moore, 2009; Rubin, 2009b, c; Rutten, 2009). After leaving the LAPD, Bratton became the chief executive at Altegrity Security Consulting, a private international security firm (Moore, 2009). Bratton is currently working at Dodgers stadium assessing security and fan services ("Dodgers," 2011; "Dodgers Hire," 2011).

[1] The scandal was named after an immigrant community where 70 LAPD officers were investigated for a pattern of corruption, including planting evidence and fabricating stories to explain unjustified shootings. As a result, in more than 100 cases, convictions were overturned or charges dropped. Despite the internal investigations into the affair, it has engendered citizen mistrust, prompting Bratton to call for an outside review in February 2003 ("Bratton," 2003).
[2] In 2002, the authorized strength was 10,196 officers (Nash, 2002).
[3] The murder rate in Los Angeles increased by 10% in 2002 over the previous year, compared to 2.3% nationally.

CAREER HIGHLIGHTS

Path-Goal Theory, Achievement-Oriented Leadership Style

Chief Richard Pennington
New Orleans Police Department (NOPD), Louisiana

In the eight years he served as Chief of NOPD, Richard Pennington addressed some of the major problems on the force, most notably with regard to police corruption.

Pennington was born in Little Rock, Arkansas, and raised in Gary, Indiana. After earning a high school degree, he enlisted in the Air Force and served in the Vietnam War. In 1968, he was discharged and subsequently took a job as a police officer in the Washington, D.C., Police Department (Stirgus, 2002). Pennington climbed the ranks of the Washington, D.C., Police Department from 1968 until 1994 (Suggs, 2002b). From 1992 to 1994, he was an assistant police chief. In the early 1990s, as a district commander, Pennington initiated the first community-policing pilot program in the department, putting more officers on foot patrols and bicycles and increasing community contact. Reportedly, between 1991 and 1992, homicides in Washington, D.C., were reduced from 137 per year to 108 (Alpert, 1994).

Pennington has a bachelor's degree in criminal justice from American University and a master's degree in criminal justice from the University of the District of Columbia. He is also a graduate of the FBI National Academy ("Richard J. Pennington," 2002).

Chief of the New Orleans Police Department (1994–2002)

When he was appointed chief of NOPD, Pennington inherited a force rife with corruption in a city losing its battle with serious crime. Pennington is credited with cutting the New Orleans murder rate in half (Perlstein, 2003). In 1994, New Orleans had 425 homicides and was considered the murder capital of the nation. By 1999, the murders had declined to 159, the largest crime-reduction

rate of the nation's 34 biggest cities (Suggs, 2002b). Murders dropped 64% between 1994 and 1999 (St. Bernard, 1999). Policies credited with this success include increasing the visibility of officers in public housing complexes, boosting the number of homicide detectives by 40%, and reducing caseloads per investigator from 12 to 8 (Varney, 1995). Pennington also raised officer pay, decentralized the detective bureau, and held commanders responsible for crime reduction (via a COMPSTAT-like focus on statistics; Perlstein, 1999c).

Pennington is also noted for addressing police corruption in New Orleans (Suggs, 2002e). The problem had been significant. At a party celebrating his inauguration as chief, an FBI agent reportedly pulled him aside to let him know that a drug ring was operating within the department. To make a break from past practice, early in his tenure Pennington replaced the Internal Affairs Department with a Public Integrity Division (PID; Varney, 1995). Two FBI agents were hired for the division in order to provide an outside perspective (Perlstein, 1999a).

Under Pennington's command, more than 350 officers were indicted, fired, or disciplined for misconduct stemming from PID investigations (Suggs, 2002b). In 1994, as aforementioned, Officer Len Davis was convicted of ordering the murder of a woman who filed a police brutality report against him (Alpert, 1998). In 1995, a veteran patrolman was charged with attempted murder for firing shots into a car filled with teenagers, two officers were charged with raping a 14-year-old while responding to a call (Perlstein, 1995), and an officer was a suspect in the killings of 24 women, although that officer was never charged and was later cleared in the killings (Perlstein, 2003). In 1998, an officer was charged with extorting money from a massage parlor (Philbin, 1998), and the second-highest-ranking officer was reprimanded for deceptive bookkeeping practices for the department's purchasing department ("High-ranking police official," 1998). In addition, in 1999, an officer was arrested for rape during a traffic stop ("Officer arrested," 1999), and another officer faced federal extortion charges for demanding money from a business owner in exchange for information about a criminal investigation against her (Perlstein, 1999b).

Another way Pennington fought corruption was by limiting the number of hours that officers could devote to moonlighting jobs, particularly since some off-duty security jobs involved criminal employers. The regulation not only reduced criminal elements in the police association but also eliminated the relevance of "power brokers," cops who controlled such off-duty hiring; it also cut down on the number of police who reported to work dangerously fatigued (Campbell, 2002).

Pennington also kept an open dialogue with police critics, such as local civil rights attorneys and the U.S. Attorney General (Stirgus, 2002). These contacts influenced Pennington to widen his war on police corruption by exploring citizen complaints of racial profiling. In 1998, he urged citizens to come forward with complaints of racial profiling so that he could investigate the alleged problem (Donze, 1998).

Recently Pennington himself was accused of wrongdoing. In January 2003, a state audit of the Southern University at New Orleans uncovered potential salary fraud in the university's compensation to Pennington, and a complaint was filed in the local district attorney's office. He was paid $44,000 over two and a half years to teach two graduate courses for five semesters, a large salary for a part-time instructor. Some signatures on Pennington's time sheets appeared to be forged. However, according to a Pennington spokesperson, he provided services in addition to teaching that included establishing a master's degree program in criminal justice and advising and recruiting students. A director of the school indicated that Pennington had indeed provided the services for which he was paid ("Ex-chief's pay," 2003; Young, 2003).

In other controversies, it was recently revealed that under Pennington's command, NOPD wrongly destroyed evidence. The current chief of the city's police department made no accusations of intentional wrongdoing on the part of Pennington but indicated that a massive evidence storage clean-up effort, which took place in 2001, may have destroyed important evidence needed in three ongoing cases (Filosa, 2003).

Upon resigning as chief in 2002, Pennington received standing ovations in New Orleans restaurants, showing the respect he had earned from many citizens (Stirgus, 2002). However, his 2002 bid for election as a Democratic mayor of New Orleans failed (Grace, 2001; Mullener, 2002). During the campaign, rumors surfaced that he abused his wife (Mullener, 2002). Furthermore, the two police unions in the city endorsed his opponent (Stirgus, 2002).

After leaving New Orleans, Pennington brought many of the same approaches to Atlanta, where he has served as chief of police since 2002. Priority issues for him include low officer morale, community policing, dangerous police practices, corruption, and police accountability.

Upon taking the job, Pennington made changes to the command staff and promoted officers in an effort to confront low morale, which he identified as one of the biggest problems with the force he inherited. Of particular note, he promoted an officer who had previously blown the whistle on the department's underreporting of crimes and had spent five years in an exiled position (Suggs, 2002e). Low morale also stemmed from complaints about low pay and a 400-member officer shortage (Niesse, 2002). Pennington announced plans to hire an additional 500 police officers in his first two years (Suggs, 2002a). Pennington also inherited a department that was deficient in necessary equipment (old motorcycle helmets and a shortage of uniforms), which he vowed to correct (Suggs, 2002c).

His inaugural promises included training new officers in community policing ("Pennington sworn in," 2002). Using New York City's quality-of-life campaign as a guide, Pennington initiated a Quality of Life Task Force to confront homelessness and panhandling in March 2003. The group was to crack down on nuisance violations, clean up litter and graffiti, and increase outreach services for homeless people. According to Pennington, panhandling, trash, and graffiti make "going downtown unpleasant, even scary" (Suggs & Ippolito, 2003, p. 1D). He has also spoken of building partnerships with the African American community, in particular, on the issue of intraracial crime (Suggs, 2003).

Pennington was clear on his expectations from the force itself. On his first day on the job, Pennington called a meeting of his new command staff and stated:

> So don't B.S. me, because I have seen it all and I have done it all. Be straightforward with me, because I can see through a lot of B.S. I am not coming here under the assumption that this is a bad police department with bad people. I like to surround myself with good people until they prove to not be capable of doing the job. Then I have no trouble replacing you. (Suggs, 2002e, p. 1E)

He also emphasized that he is a uniformed police chief because it sends the message that he is part of the department and because it is a sign of authority (Suggs, 2002e).

In November 2002, Pennington reformed the department's policy on high-speed chases. In the previous six months, there had been five fatalities after high-speed chases, and only one of the chases stemmed from a serious crime. Under the new policy, chases are only permitted for serious offenses against people, such as murder, rape, or robbery. Pennington implemented a similar policy as chief in New Orleans, and the numbers of high-speed chase injuries and fatalities were subsequently reduced (Suggs, 2002f).

In recent controversies, the Atlanta Police Department shot 12 people (5 fatally) between January and August 2002, more than in any of the previous six years. The press reported that community mistrust of officers was on the rise as a result. Some community members accused the department of racial profiling, particularly of African Americans driving expensive cars, which described at least one of the shooting victims. Pennington responded by suspending undercover vice operations and ordering a review of department policies on the use of lethal force. One of the shootings involved a cop shooting at a car that was veering toward him at a high speed, which prompted Pennington to order a specific review of the policy regarding shooting at cars (such shootings were prohibited when Pennington headed the New Orleans police). He also stated, however, that the department did not have a history of problems with excessive use of force ("Atlanta suspends undercover vice unit," 2002; Niesse, 2002). Pennington indicated that he was open to a federal investigation into the shootings if that would help resolve the controversy as to what happened and the appropriateness of the police response. He also stated, "I just want my officers to know that we are going to thoroughly investigate shootings. If they have not done anything improper, they have nothing to worry about" (Suggs, 2002d, p. 3D).

Moreover, Pennington has advocated for a rejuvenated civilian review board (although Atlanta has one, it was not active in 2002). This commitment to civilian review was praised by an editorial in *The Atlanta Journal and Constitution* as an important check against a perceived blue wall of silence (Tucker, 2002).

Pennington exhibits the achievement-oriented approach to the path-goal theory. The problems he faced when taking on the position in New Orleans were even more serious in nature than the ones faced by Commissioner Bratton. The level of corruption of the NOPD was absolutely unprecedented in the history of modern policing in the United States. The cleaning up of the ranks, changes in the training procedures, pay raises, establishment of new mechanisms of accountability—all were far-reaching steps in a major attempt not only to clean up the department that was often referred to as "the most corrupt police department in the United States" but also to elevate it to a police department that would become a model for others.

Chief Pennington came to New Orleans with an ambitious plan. Its success was dependent on the effective motivation of those officers who remained on the force, so they would be an asset to build on in the creation of a new environment while breaking away from the old image and subculture. He, as well as Bratton in New York City, had to buy both the minds and hearts of his officers to achieve this reengineering. Establishing high standards definitely has its merits in environments of highly questionable integrity and scarce resources for change. Chief Pennington was not a man of incremental steps, which would have been more realistic; his changes instead were extreme, swift, and innovative. However, he was not able to create the intellectual and emotional safety needed to successfully "purchase" the minds and hearts of those who needed to maintain the high standard of excellence that the chief demanded. When he ran for mayor, police unions endorsed his opponent.

One of his biggest problems in not being able to create this intellectual and emotional safety was his inability to change the force in the most extreme way by purging the force of those officers who needed to be fired. At a certain point, many officers were charged with criminal misconduct and subsequently fired from the force. Although these officers were charged and found guilty, many more officers witnessed these officers' misconduct yet remained on the force and were never charged or disciplined. With so many officers who had been contaminated by the previous misconduct and wrong practices remaining on the force, it was impossible for Chief Pennington to buy enough minds and hearts to enable the commitment and motivation to sustain the changes he ultimately sought.

In November 2006, Pennington had to deal with the fallout of three Atlanta police officers being shot by 88-year-old Kathryn Johnston as they served a no-knock warrant at her residence (Gross, 2006; Halnes, 2006; Serrano, 2006). Johnston was shot and killed by police during the encounter (Gross, 2006; Halnes, 2006; Serrano, 2006). In response to the incident, Pennington ordered his department to review its policy on the use of no-knock warrants and the use of confidential informants (Gross, 2006; Halnes, 2006; Serrano, 2006).

About a week after the incident, the informant whose testimony the warrant to search Johnston's residence was based on, came out and denied ever buying drugs from Johnston's home (Haines, 2006; "Informant," 2006). According to the informant, the police officers asked him to go along with a story which was concocted by the police officers after the shooting had occurred ("Informant: Police," 2006). That story involved him telling police that he had brought drugs at the Johnston residence from a man named Sam ("Informant: Police," 2006).

In response to the controversial nature of the shooting, the FBI launched an investigation into the incident on November 28, 2006 ("Mourners," 2006; Turner, 2006). Pennington also placed seven narcotics investigators and one sergeant on paid leave pending the outcome of the investigation (Dewan & Goodman, 2006; "Mourners," 2006; Turner, 2006).

The incident also had a monetary impact on Pennington because as a result of the shooting, the city council said no to increasing his annual pension plan contribution by $10,000, a move supported by the council prior to the shooting ("Council," 2006).

A year later, in November 2007, Pennington changed department policy on no-knock search warrants stating that from now on, no-knock warrants must be approved by a major and the officers executing those warrant must wear special uniforms ("Atlanta," 2007). Pennington also replaces all 15 members of the narcotics squad with 30 newly trained officers ("Atlanta," 2007; "Narc," 2007).

In 2008, Pennington had to deal with two officers being wounded and the suspect killed during a shootout ("2 Officers hurt," 2008; "Police Kill," 2008). The incident is notable because the officers were part of a unit whose goal was to decrease burglaries in southwest Atlanta, one of Pennington's initiatives to reduce crime in Atlanta ("2 Officers hurt," 2008; "Police Kill," 2008).

On July 31, 2009, in the wake of several high profile shootings, Pennington announced that 27 new police officers would hit the streets in areas with carjacking problems ("4 Killed," 2009). Pennington also announced that 139 new officers would be deployed by year's end in an effort to reduce crime ("4 Killed," 2009).

Six weeks later, on September 14 2009, Pennington had to deal with the fallout of the raid at Atlanta Eagle, a gay bar (Rankin, 2009). According to Pennington, vice officers had been to Atlanta Eagle on two prior occasions and witnessed illegal activities (Rankin, 2009). During the raid, eight employees were arrested on improper permit charges (Rankin, 2009). The incident was controversial because according to eyewitnesses, officers ordered patrons in the bar to lie face down on the floor (Rankin, 2009). It was also alleged that officers illegally searched the patrons and taunted them with anti-gay slurs (Rankin, 2009). Pennington disagreed with critics accusations, saying that the patrons were not illegally searched but instead frisked by police officers to ensure those officers safety (Rankin, 2009).

On November 2, 2009, Pennington announced that he would resign at the end of the year ("Atlanta Police Chief," 2009; Boone, 2009). Shortly after resigning, he suffered a stroke on Memorial Day, in 2010 (Cook, 2010).

Exercise

The police commissioner and chief detailed in this chapter have been examined through the prism of path-goal theory. Using what you have learned from other theories in this book or from the summaries (as indicated in the Introduction) you have read thus far, it is possible to consider their approaches through the lens of different theories. What other theories can you find applicable to the commissioner and the chief? Are there lessons to be learned from those applications, or do those theories indicate alternative paths or approaches that might have yielded a more successful outcome?

References

A test for Commissioner Bratton. (1993, October 2). *The Boston Globe*, p. 10.

Alpert, B. (1994, October 15). Community patrols big with new chief. *The Times-Picayune*, p. A1.

_____(1998, July 8). Rights group praises Pennington's efforts, but brutality problems persist, it claims. *The Times-Picayune*, p. B8.

Atlanta police chief announces resignation. (2009, November 25). *The Augusta Chronicle*. Retrieved June 6, 2011, from http://chronicle.augusta.com/stories/2009/11/25/met_557032.shtml.

Atlanta police overhaul raid procedures. (2007, November 21). *UPI*. Retrieved June 6, 2011, from http://www.upi.com/Top_News/2007/11/21/Atlanta-police-overhaul-raid-procedures/UPI-39471195659548/.

Atlanta suspends undercover vice unit after fatal shooting. (2002, August 6). *The Associated Press*, state and local wire.

Blood, M. (1995, June 10). NYPD vows to improve tarnished reputation. *The Record*, p. A4.

Boone, C. (2009, November 24). Pennington makes it official, announces resignation. *The Atlanta Journal-Constitution*. Retrieved June 6, 2011, from http://www.ajc.com/news/atlanta/pennington-makes-it-official-212899.html.

Bratton calls for independent Rampart probe. (2003, February 28). Retrieved June 22, 2003, from http://www.kcal9.com/topstories/topstoriesla_story05919851.html.

Broder, J.M. (2003, June 16). Los Angeles city council retaliates against mayor: Fracas over police and rejection of budget. *The New York Times*, p. A12.

Bryman, A. (1986). *Leadership and organizations.* London: Routledge & Kegan Paul.

Campbell, C. (2002, July 23). Police problems need action from new chief. *The Atlanta Journal and Constitution,* p. 2B.

Celebs urge legal curbs on aggressive paparazzi. (2008, August 1). *Malaysia Star.* Retrieved June 6, 2011, from http://thestar.com.my/news/story.asp?file=/2008/8/1/apworld/20080801101406&sec=apworld.

Community policing, Bratton style. (1994, January 31). *The New York Times,* p. A16.

Cook, R. (2010, June 17). Ex-APD Chief Richard Pennington has a stroke. *Atlanta Journal-Constitution.* Retrieved June 6, 2011, from http://www.ajc.com/news/atlanta/ex-apd-chief-richard-551001.html.

Council votes against pension increase for Atlanta police chief. (2006, December 5). *Elizabethon Star.* Retrieved June 6, 2011, from http://news.google.com/newspapers?id=IrljAAAAIBAJ&sjid=ArMMAAAAIBAJ&pg=1420,3811858&dq=chief+richard+pennington&hl=en.

Dessler, G. (1972). A test of a path-goal motivational theory of leadership. *Proceedings of Academy of Management,* 178–181.

Dewan, S. & B. Goodman (2006, November 28). Atlanta officers suspended in inquiry on killing in raid. *New York Times.* Retrieved June 6, 2011, from http://www.nytimes.com/2006/11/28/us/28atlanta.html.

Dodgers hire former LAPD chief for security review. (2011, April 7). *ABC.* Retrieved June 6, 2011, from http://abc-news.go.com/Sports/wireStory?id=13314207.

Dodgers hire William Bratton to assess stadium security. (2011, April 6). *Los Angeles Daily News.* Retrieved June 6, 2011, from http://www.dailynews.com/rss/ci_17786582.

Don't mention the war: A delicate task for William Bratton in Los Angeles. (2003, March 22). *Economist,* 366 (8316).

Donze, F. (1998, November 17). Police behavior overtakes budget. Won't tolerate misconduct, Pennington vows. *The Times-Picayune,* p. B1.

Dussault, R. (1999, August). The taking of New York: Former NYC police commissioner William Bratton talks about how technology helped make the streets safer. *Government Technology.* Retrieved June 22, 2003, from www.govtech.net/magazine/gt/1999/aug/bratton-fldr/bratton/phtml.

Ex-chief's pay target of investigation. (2003, January 15). *The Associated Press,* state and local wire.

FBI to probe LA police clash at immigrant rally. (2007, May 3). *Reuters.* Retrieved June 6, 2011, from http://www.reuters.com/article/2007/05/04/idUSN03267874.

Filosa, G. (2003, February 6). Evidence lost, police chief concedes. *The Times-Picayune,* p. 1.

Geis, S. (2007, May 8). Top LAPD officers disciplined for use of force at rally. *Washington Post.* Retrieved June 6, 2011, from http://www.washingtonpost.com/wp-dyn/content/article/2007/05/07/AR2007050701640.html?hpid=sec-nation.

George, T. (1995, September 3). Bratton's new breed of blue. *Daily News,* p. 1.

Georgopoulos, B.S., G.M. Mahoney & N.W. Jones, Jr., (1957). A path-goal approach to productivity. *Journal of Applied Psychology,* 41, 345–353.

Grace, S. (2001, November 21). Pennington launches campaign for mayor. *The Times-Picayune,* p. 1.

Greene, J.A. (1999). Zero tolerance: Police policies and practices in New York City. *Crime and Delinquency,* 45 (2), 171–187.

Gross, D. (2006, November 27). FBI to probe elderly woman's shooting. *Fox News.* Retrieved June 6, 2011, from http://www.foxnews.com/wires/2006Nov27/0,4670,ElderlyShootout,00.html.

Haines, E. (2006, December 4). Sharpton wants congressional inquiry of Atlanta shooting. *The Tuscaloosa News.* Retrieved June 6, 2011, from http://news.google.com/newspapers?id=zHUjAAAAIBAJ&sjid=acEAAAAIBAJ&pg=5219,3112203&dq=chief+richard+pennington&hl=en.

Halnes, E. (2006, December 5). Officials meet with residents in troubled neighborhood. *Rome News-Tribune.* Retrieved June 6, 2011, from http://news.google.com/newspapers?id=JuQvAAAAIBAJ&sjid=bzsDAAAAIBAJ&pg=3495,2712273&dq=chief+richard+pennington&hl=en.

Helfand, D., P. McGreevy & A. Blankstein (2007, May 8). LAPD officers to be reassigned for role in response to rally. *Pittsburgh Post-Gazette.* Retrieved June 6, 2011, from http://news.google.com/newspapers?id=FlRAAAAIBAJ&sjid=GnIDAAAAIBAJ&pg=2723,2514199&dq=chief+william+j+bratton&hl=en.

High-ranking police official gets reprimand, suspension. (1998, September 17). *The Associated Press,* state and local wire.

House, R.J. (1996). Path-goal theory of leadership: Lessons, legacy, and a reformulated theory. *Leadership Quarterly,* 7 (3), 323–352.

_____ (1971). A path-goal theory of leader effectiveness. *Administrative Science Quarterly,* 16, 321–328.

_____ & G. Dessler (1974). Path-goal theory of leadership. *Journal of Contemporary Business,* 3, 81–97.

Indvik, J. (1986). Path-goal theory of leadership: A meta-analysis. *Proceedings of Academy of Management,* 189–192.

Informant denies buying drugs at elderly Atlantan's home. (2006, November 28). *CNN.* Retrieved June 6, 2011, from http://articles.cnn.com/2006-11-28/us/atlanta.shooting_1_informant-kathryn-johnston-drug-raid?_s=PM:US.

Informant: Police lied about shooting. (2006, November 28). *St. Petersburg Times.* Retrieved June 6, 2011, from http://news.google.com/newspapers?id=8DxSAAAAIBAJ&sjid=nHgDAAAAIBAJ&pg=5073,2008267&dq=chief+richard+pennington&hl=en.

Jermier, J. (1996). The path-goal theory of leadership: A subtextual analysis. *Leadership Quarterly,* 7 (3), 311–316.

Kandel, J. (2002, November 21). Dial L.A. for murder: Movie capital leads nation in homicides. *The Daily News of Los Angeles,* p. N1.

4 Killed overnight in metro Atlanta. (2009, July 31). *WSB.* Retrieved June 6, 2011, from http://www.wsbtv.com/news/20234521/detail.html.

Laird, L. (2002, October 4). Bratton pledges to restore LAPD's "tarnished" reputation as police chief. *Metropolitan News Enterprise,* p. 8.

Lakshmanan, I. (1993, June 30). Community leaders say Bratton able, ambitious, and some officers express optimism. *The Boston Globe,* p. 15.

L.A. Mayor returning to deal with rally clash. (2007, May 4). *USA Today.* Retrieved June 6, 2011, from http://www.usatoday.com/news/nation/2007-05-04-immigration-rally_N.htm.

LA officer held in cuffed teen assault. (2006, December 7). *Fox News.* Retrieved June 6, 2011, from http://www.foxnews.com/wires/2006Dec07/0,4670,PoliceAssault,00.html.

LA police arrest cop over taped assault on teen. (2006, December 8). *Seattle Times.* Retrieved June 6, 2011, from http://seattletimes.nwsource.com/html/nationworld/2003467340_lapd08.html.

LA police dismiss paparazzi plans. (2008, August 1). *BBC.* Retrieved June 6, 2011, from http://news.bbc.co.uk/2/hi/entertainment/7537048.stm.

LA to tame paparazzi wolves. (2008, August 2). *The Australian.* Retrieved June 6, 2011, from http://www.theaustralian.com.au/news/la-to-tame-paparazzi-wolves/story-e6frg6tf-1111117083552.

Leovy, J. (2003, January 1). L.A. crime edges up, led by 10% jump in killings. *The Los Angeles Times,* p. B6.

Lewis, D.L. & C. Siemaszko (1995, July 12). Bratton blows top. *Daily News,* p. 4.

Los Angeles police chief promises investigation into immigration rally clash. (2007, May 3). *The China Post.* Retrieved June 6, 2011, from http://www.chinapost.com.tw/latestnews/200753/45904.htm.

Los Angeles Police Chief William Bratton says officers used inappropriate tactics during immigration rally. (2007, May 2). *Fox News.* Retrieved June 6, 2011, from http://www.foxnews.com/story/0,2933,269616,00.html.

Marzulli, J. (1995, November 8). Bratton sez judges killing city's quality of life. *Daily News,* p. 2.

Mathieu, J.E. (1990). A test of subordinates' achievement and affiliation needs as moderators of leader path-goal relationships. *Basic and Applied Psychology,* 11 (2), 179–180.

McGrory, B. (1993, June 30). Bratton seen as a leader in "new school" of policing. *The Boston Globe,* p. 15.

McQuillan, A. (1995, August 24). Gripes vs. cops skyrocket under independent CCRB. *Daily News,* p. 36.

Moore, S. (2009, August 5). Los Angeles chief to step down. *New York Times.* Retrieved June 6, 2011, from http://www.nytimes.com/2009/08/06/us/06bratton.html.

Mourners remember elderly shootout victim. (2006, November 29). *Sun Journal.* Retrieved June 6, 2011, from http://news.google.com/newspapers?id=ZuUpAAAAIBAJ&sjid=aWQFAAAAIBAJ&pg=3928,4877076&dq=chief+richard+pennington&hl=en.

Mr. Bratton's wise policing. (1994, May 31). *The New York Times,* p. A30.

Mullener, E. (2002, May 24). Farewell to the chief. *The Times-Picayune,* p. 1.

Narc unit replaced after woman, 92, killed. (2007, May 23). *USA Today.* Retrieved June 6, 2011, from http://www.usatoday.com/news/nation/2007-05-23-elderly-shooting_N.htm.

Nash, J. (2002, November 20). Bratton backs off 12,000 goal: Cash-strapped city to add 270 to LAPD. *The Daily News of Los Angeles,* p. N3.

New Boston police chief outlines his priorities for the department. (1993, July 11). *The Boston Globe,* p. 2.

Newfield, J., & Jacobson, M. (2000, July/August). An interview with William Bratton. *Tikkun,* 15 (1), 51–54.

Niesse, M. (2002, August 21). Killings by Atlanta police raise racial tensions. *Chattanooga Times Free Press,* p. NG4.

Northouse, P.G. (2004). *Leadership: Theory and practice* (3rd ed.). Thousand Oaks, CA: Sage.

NYPD, Inc. (1995, July 29). *Economist,* 336 (7925), 50.

Officer arrested in sexual assaults. (1999, August 26). *The Advocate,* p. B6.

2 Officers hurt, suspect killed in Ga. shootout. (2008, July 15). *USA Today.* Retrieved June 6, 2011, from http://www.usatoday.com/news/nation/2008-07-15-atlanta_N.htm.

Pennington sworn in as new Atlanta police chief. (2002, July 9). *The Associated Press,* state and local wire.

Perlstein, M. (1995, May 23). 2 N.O. cops booked with raping 14-year-old. *The Times-Picayune,* p. B1.

_____(1999a, July 26). Reformed NOPD is hot topic globally: Its top cop active on lecture circuit. *The Times-Picayune,* p. B1.

_____(1999b, October 2). N.O. cop booked in extortion scheme: FBI says cop traded information for cash. *The Times-Picayune,* p. B3.

_____(1999c, October 15). Murder rate nearly half, chief says. *The Times-Picayune,* p. B1.

_____ (2003, May 29). Manhunt thrusts Pennington back into N.O. spotlight. *The Times-Picayune*, p. 1.

Philbin, W. (1998, April 15). Cop held in massage parlor sting. *The Times-Picayune*, p. A1.

Police kill burglary suspect in shootout. (2008, July 16). *The Albany Herald*. Retrieved June 6, 2011, from http://news.google.com/newspapers?id=US1EAAAAIBAJ&sjid=RrEMAAAAIBAJ&pg=2803,1961104&dq=chief+richard+pennin gton&hl=en.

Prengaman, P. (2007a, May 3). Immigrant groups decry police tactics. *Boston Globe*. Retrieved June 6, 2011, from http://articles.boston.com/2007-05-03/news/29235813_1_rubber-bullets-batons-officers.

Prengaman, P. (2007b, May 3). Police to review clash at LA rally. *Washington Post*. Retrieved June 6, 2011, from http://www.washingtonpost.com/wp-dyn/content/article/2007/05/02/AR2007050201705.html.

Rankin, B. (2009, September 14). Chief: Vice saw sex at gay bar. *The Atlanta Journal-Constitution*. Retrieved June 6, 2011, from http://www.ajc.com/news/atlanta/chief-vice-cops-saw-138328.html.

Richard J. Pennington, chief of police. (2002, July 29). Retrieved June 9, 2003, from www.atlantapd.org/cstaff/chief/html.

Richards, D. (2002). *Convincing minds and winning hearts*. Louisville, KY: Brown Herron Publishing.

Rubin, J. (2009a, June 5). LAPD Chief William J. Bratton awarded major British title. *Los Angeles Times*. Retrieved June 6, 2011, from http://articles.latimes.com/2009/jun/05/local/me-bratton5.

_____ (2009b, August 5). Los Angeles police chief William J. Bratton to step down. *Los Angeles Times*. Retrieved June 6, 2011, from http://latimesblogs.latimes.com/lanow/2009/08/los-angeles-police-chief-william-bratton.html.

_____ (2009c, August 6). William Bratton announces he will resign as LAPD chief. *Los Angeles Times*. Retrieved June 6, 2011, from http://articles.latimes.com/2009/aug/06/local/me-bratton6.

_____ & Zahniser, D. (2009, October 10). Bratton is still firm about the size of the force. *Los Angeles Times*. Retrieved June 6, 2011, from http://articles.latimes.com/2009/oct/10/local/me-bratton10.

Rutten, T. (2009, October 17). Chief Bratton's too-brash exit. *Los Angeles Times*. Retrieved June 6, 2011, from http://articles.latimes.com/2009/oct/17/opinion/oe-rutten17.

Schriesheim, C.A. & M.A. Von Glinow (1977). The path-goal theory of leadership: A theoretical and empirical analysis. *Academy of Management Journal, 20*, 398–405.

Serrano, A. (2006, November 27). Atlanta police to review no-knock policy. *CBS*. Retrieved June 6, 2011, from http://www.cbsnews.com/stories/2006/11/27/national/main2210590.shtml.

Stars ask for more protection from photogs. (2008, August 1). *UPI*. Retrieved June 6, 2011, from http://www.upi.com/Entertainment_News/2008/08/01/Stars_ask_for_more_protection_from_photogs/UPI-67791217595425/.

St. Bernard, S.C. (1999, August 6). N.O. police chief shares inspiration at breakfast: Pennington gets spiritual on crime. *The Times-Picayune*, p. B1.

Steinhauer, J. & J. Preston (2007, May 4). Action by police at rally troubles Los Angeles chief. *New York Times*. Retrieved June 6, 2011, from http://www.nytimes.com/2007/05/04/us/04immig.html.

Stirgus, E. (2002, April 14). New Orleans chief enters city's folklore. *Cox News Service,* Domestic.

Stockstill, M. (2002, October 29). Bratton lays out priorities for LAPD command staff. *City News Service*.

Suggs, E. (2002a, May 31). Pennington names Atlanta's top cop. *The Atlanta Journal and Constitution*, p. 1C.

_____ (2002b, June 6). New chief wants raise for officers. *The Atlanta Journal and Constitution*, p. 1JN.

_____ (2002c, July 13). Inside the mind of the new chief. *The Atlanta Journal and Constitution*, p. 1H.

_____ (2002d, July 26). Atlanta chief welcomes U.S. probe of shooting. *The Atlanta Journal and Constitution*, p. 3D.

_____ (2002e, October 19). Chief shifts police brass. *The Atlanta Journal and Constitution*, p. 1E.

_____ (2002f, November 8). Deadly high-speed chases. *The Atlanta Journal and Constitution*, p. 1C.

_____ (2003, May 23). Area chief joins efforts to fight black crime rate. *The Atlanta Journal and Constitution*, p. 2F.

_____ & M. Ippolito (2003, March 27). Mayor, police chief focus on homeless. *The Atlanta Journal and Constitution*, p. 1D.

Tucker, C. (2002, July 17). Cops needn't fear scrutiny of civilians. *The Atlanta Journal and Constitution*, p. 16A.

Turner, D. (2006, November 28). Service for woman killed in cop shootout. *Washington Post*. Retrieved June 6, 2011, from http://www.washingtonpost.com/wp-dyn/content/article/2006/11/28/AR2006112800822.html.

Varney, J. (1995, June 30). Department cleaning up act, Pennington tells commission. *The Times-Picayune*, p. B1.

Vroom, V.H. (1964). *Work and motivation*. New York: Wiley.

Walker, S. & C.M. Katz (2002). *The police in America* (4th ed.). New York: McGraw-Hill.

Watkins, T. (2008, August 3). LA chiefs ready to snap over paparazzi. *Scotsman*. Retrieved June 6, 2011, from http://news.scotsman.com/world/LA-chiefs-ready-to-snap.4352504.jp.

Wilborn, P. (2002, October 3). With Bratton LA lands a star for its police chief. *The Associated Press,* state and local wire.

William J. Bratton, chief of police. (n.d.). Retrieved August 10, 2004, from http://www.lapdonline.org/c_o_s/bratton_w_bio.htm.

Wilson, J.Q. & G.L. Kelling (2011). "Broken Windows." In Francis T. Cullen & Robert Agnew (Eds.), *Criminological Theory: Past to Present* (4th ed., pp.189–197). New York: Oxford University Press.

Wofford, J.C. & L.Z. Liska (1993). Path-goal theories of leadership: A meta-analysis. *Journal of Management,* 19 (4), 857–876.

Young, T. (2003, January 14). Auditor questions ex-chief's SUNO pay. *The Times-Picayune,* p. 1.

Yukl, G. (1994). *Leadership in organizations* (3rd ed.). Englewood Cliffs, NJ: Prentice Hall.

10

Leadership and Command of the Critical Incident
Psychodynamic Approach

OVERVIEW OF THE APPROACH

The psychodynamic approach to leadership focuses on the leader as a charismatic individual stating that deep-seated emotional factors, such as lifetime experiences, affect leadership behavior (Stech, 2004). This approach is examined via the example of New York Police Department (NYPD) Commissioner Bernard Kerik, who oversaw the NYPD through the September 11, 2001, terrorist attacks.

According to Max Weber, there are three bases of legitimate authority: traditional, legal, and charismatic; traditional represents the followers' tendency to conform based on ancient patriarchal cultures; legal rests in the acceptance of the rationality of rules made in an organization/society; and charismatic resides in individuals who possess confidence and heroism (e.g., prophets) (Weber, 1964).

People obey leaders because they believe in them, a phenomenon present in all places and historical epochs (Weber, 1973). Charisma arises from followers experiencing distress, which sparks devotion to the leader (Weber, 1973).

In *Group Psychology and the Analysis of the Ego* (1921), Freud describes group needs: continuity, definition, interaction (with other groups), and traditions, which counter a primitive state characterized

---

Three Bases of Legitimate Authority

Traditional
- Habitual orientation based on ancient patriarchal culture

Legal
- Acceptance of the rationality of rules

Charismatic
- Authority of individuals with personal confidence and heroism, such as prophets

(Weber, 1964)

by intellectual disintegration and declining consciences. In *Moses and Monotheism* (1939), Freud describes the beginning of civilization as a struggle with the leader of a primitive clan, reciting a myth that the primal father of the first group was the target of envy and was slain by his son; removing the object of the love of the father, resulting in melancholic object loss for the son; a void filled by loving a new leader.

Consequently, groups require organization and leadership to counter negative unconscious desires and narcissism. Freud posits that libido (i.e., energy necessary for love) links subordinates to leaders; likewise, one follows a leader out of love (Freud, 1921; Winer et al., 1984–1985). Charisma is a type of love where the leader represents a father figure, and a source of love and fear. Followers believe in the leader, who is a surrogate for the father whose image they incorporate into their superego.

Kohut (1976), states that charismatic leaders experience trauma, propelling them to use leadership to bolster their self-esteem. Subordinates, meanwhile, cope with their shame, jealousy, and hate by idolizing leaders (Bass, 1990).

LaBarre (1972) posits that subordinates' fascination with leaders stems from leaders' narcissistic fascination with themselves; a fascination derived from the fascination that their mother had for them. Leaders emerge in crisis situations when cultural norms fail. Kets de Vries (1980) indicates that leader's grandioseness emerges during crises. Simultaneously, subordinates feel helpless, which draws the two together. Leaders apply their paranoid potential to the problem facilitating effective leadership.

Winer et al. (1984–1985) interprets the psychoanalytic approach to mean that an individual's suffering facilitates self-transformation: a process by which individuals gain the capacity to lead others. Charismatic leaders, after suffering a trauma, unconsciously feel passive, "powerless," and acted upon by the object, despite being the "doer" rather than "the done to" (Winer et al., 1984–1985, p. 169). Followers have an unconscious sense of self, attracting the leader to the followers as a paradigmatic reenactment of the original trauma:

> In the unconscious fantasy, [the leader] remains passive and helpless despite success after success. . . . The leader seeks followers in order to immerse himself repeatedly in the traumatic situation and repeatedly reverse it both for himself and others. (Winer et al., 1984–1985, p. 170)

However, the psychoanalytic approach does not explain the charismatic behavior of healthy leaders (Strozier & Offer, 1985). Zaleznik (1974) recognizes that some leaders have a healthy development of self, related to the psychoanalytic concept of "introjection." When individuals internalize the mother/father figure in childhood (which occurs in stable, emotionally healthy families), they can draw on that internalization in times of object loss. Basically, they have introjected their objects of needs and are not propelled to crisis. Zaleznik (1977) believes that charismatic leaders have introjected objects: the basis for their ties to followers. An internal audience (mother or father) forms the leaders' inner life, helping them to negotiate difficulties on behalf of others.

In exploring self-actualization in the workplace, Whyte (1994) writes:

> Once basic necessities are taken care of, there are other more immediate urgencies central to human experience, and it is these urgencies that are continually breaking through our fondest hopes for an ordered work life. The split between what is nourishing at work and what is agonizing is the very chasm from which our personal

destiny emerges. Accepting the presence of this chasm we can begin to deal, one step at a time, with the continually hidden, underground force that shape our lives, often against our will. (p. 5)

Similarly, Jungian psychology focuses on the shadow self—the part of the psyche that is denied because it is considered unacceptable (Stech, 2004). Whyte encourages workplace leaders to become aware of their shadow selves, described as unresolved forces that affect their behavior. Without awareness, leaders unconsciously play out the same struggles and cannot be fulfilled or effective.

BIRTH ORDER

Adler (1928) states that intellectual development and the need for power is more prevalent among firstborns who are overrepresented among high ranking politicians across the globe, leading researchers to believe that there is a relationship between birth order and leadership. One hypothesis is that firstborns initially receive all their parents' attention and resources. Firstborns also are leaders to their siblings, followers of their parents, and participants in sibling rivalry, which prepares them for leadership roles as adults and the struggle associated with those roles (Andeweg & Van Den Berg, 2003).

In more Freudian terms, Adler (1928) refers to the "dethronement" of firstborns when siblings are born, which sparks a crisis of deprivation (compare this to Freud's notion of object loss), resulting in the need to seek power in order to compensate. Andeweg and Van Den Berg (2003) conducted a survey of Dutch politicians about birth order confirming the overrepresentation of firstborns in leadership positions and that gender (being female) does not mitigate the effects of birth order.

TRANSACTIONAL ANALYSIS

According to Berne (1961), there are three ego stages (parent, adult, and child). Leaders should attempt to develop into an adult ego state characterized by being able to assess situations and develop solutions. Management consultants at McKinsey & Co. explored training employees in transactional analysis. According to their model, managers and employees desired to grow and wanted a deeper emotional connection to their work. Role-playing and reflection enabled

---

Birth Order

- Intellectual development and a need for power are more prevalent in firstborn children.
- An indirect relation exists between birth order and leadership.
- Firstborns are differentially parented in that they initially receive all their parents' attention and resources.
- Firstborns have a dual experience: leaders of their younger siblings and followers of their parents.
- Unlike singletons, firstborns experience sibling rivalry, preparing them for the political struggle of being a leader.
- Dethronement—when new siblings are born—causes in firstborns a crisis of deprivation, from which emerges the need to seek power to compensate.

individuals to "tap into their rational and subconscious hope for the future" (Bellin & Rennie, 2003, p. 37). This growth is similar to Berne's (1961) notion of maturing into an adult ego state in order to maximize cognitive abilities considered connected to psychoemotional needs.

PSYCHOHISTORY

Psychohistorians attempt to explain political leaders via the psychoanalytic approach. Freud and Bullitt analyzing the relationship between Woodrow Wilson and his father found that Wilson idealized his father to the point of obsession, resenting his father's authority while admiring his articulateness, resulting in Wilson's political ambition to undertake projects such as the League of Nations (Bass, 1990). George and George (1998) elaborated on what led to Wilson's idealization and resentment of his father.

PERSONALITY AND LEADERSHIP

Psychologists study leadership using personality instruments such as the Myers-Briggs Personality Profile (Stech, 2004). Kroeger and Theusen (1992) proposed that knowing the personality profiles of followers, or "typewatching," allows leaders to be more effective. This approach proposes that a leader's awareness of his/her type and that of his/her follower's leads to better communication, productivity, time and stress management, and conflict resolution.

Mann (1959) performed a meta-analysis of 28 studies concerning the relation between personality and evaluations of leadership in small groups; finding that measures of intelligence, adjustment, extroversion, and dominance are positively related to effective leadership.

More recently, psychologists considered leadership via the big five model of personality, which states that personality can be described along five dimensions—"surgency" (extraversion), agreeableness, conscientiousness, emotional stability, and intellect. Leaders decide who their ideal subordinate is based on these personality dimensions and build their team accordingly (Hogan, Curphy, & Hogan, 1994).

Zaleznik (1977) contrasts managers and leaders based on personality characteristics. Managers are short-term thinkers, embedded in routine and inflexibility; while leaders possess characteristics to develop ideas for long-standing problems. Zaleznik (1977) emphasizes talent over learned skills; however, individuals with talent need mentoring to actualize their potential. "The only sure way an individual can interrupt reverie-like preoccupation and self-absorption is to form a deep attachment to a great individual or other person who understands and has the ability to communicate with the gifted individual" (Zaleznik, 1977, p. 75).

"Typewatching"

- Being aware of the personality profiles of followers can allow the leader to be more effective.
- This approach does not condone any particular personality type as being more conducive to leadership.
- The leader's awareness of his/her type and that of followers may result in better communication, productivity, time management, stress management, and conflict resolution.

(Kroeger & Theusen, 1992)

Effective Leadership

Behaving as and becoming an effective leader is a secondary by-product of an intense commitment to a purpose.

PSYCHODYNAMIC APPROACH TO LEADERSHIP AND THE CASE OF COMMISSIONER BERNARD KERIK

The psychodynamic approach appears to be the most straightforward to explain behaviors of certain leaders, providing that enough information exists to explain and understand the behaviors described. This type of leadership is associated with the individual's personality and self-insight which allows them to better understand the situation and the people involved.

The question is whether such individuals are capable of using this knowledge and understanding to manipulate subordinates verses simply working with them. Does awareness of one's "dark side" translate into control over it, in an attempt to avoid errors of judgment? Or will someone, aware of his/her inner struggle and inner demons justify the manipulation of subordinates by using the existence of this "dark side" as an excuse?

Whatever the answers to these questions, it is a deeply emotional approach to leadership. In police environments, emotions should be kept in check to maximize effort. No other situation/event in history has called for more self-restraint and the need to keep emotions in check than 9/11: a disaster that Commissioner Bernard Kerik oversaw the NYPD response to it. Although the media and the NYPD indicate that the department was prepared for a disaster like 9/11, there is no doubt that without a competent leader that the response would have fallen apart. Kerik's competence has many roots and origins; however, this author believes his childhood experiences enabled him to unite the forces, preventing this horrible disaster from becoming even more traumatic. Of course, this is only one incident, and only one perspective on his leadership during that incident, but it is an excellent illustration of how the psychodynamic approach can apply in a policing context.

CAREER HIGHLIGHTS

Psychodynamic Approach

Commissioner Bernard Kerik
New York Police Department, New York

Bernard Kerik was born on September 4, 1955, in Newark, New Jersey. His parents separated in 1957, and he moved with his mother to Ohio where she abandoned him at the age of three, leaving him in the care of her boyfriend's mother (Kerik, 2001; Marzulli, 2001b). After abandoning Kerik, his mother worked as a prostitute and was killed in 1964. In 1959, his father regained custody of him and brought him back to Newark, where he grew up under the care of his aunt Clara (Kerik, 2001). Kerik dropped out of high school; however, his training in martial arts as a teenager helped him to segue into a military career (Kerik, 2001). In 1974, Kerik joined the U.S. Army and was stationed in

Korea in 1975, where he fathered a daughter whom he lost track of after 1978 (Kerik, 2001). After investigating his daughter's whereabouts extensively, Kerik was reunited with her in 2002 after her mother had watched Kerik on the *Oprah Winfrey Show* in 2001 ("Former police commissioner," 2002).

After being discharged from the Army in 1977, Kerik worked for the Interstate Revenue Research Center investigating cigarette smugglers in South Florida. From 1978 to 1981, he worked as a military security officer in Saudi Arabia; protecting American architects, engineers, and construction crews building a military base near the Iraq border. Kerik then became a Passaic County, New Jersey, corrections officer. In 1982, he returned to Saudi Arabia for a second tour, which lasted until 1984. In 1985, Kerik became assistant commander of the Sheriff's Emergency Response Team (SERT) in Passaic County. In 1987, he joined the NYPD and had a career as a narcotics task force officer before rising through the ranks.

From 1998 to 2000, Kerik was New York City's corrections commissioner (Pyle, 2001). He became the NYPD police commissioner in August 2000. During his tenure, serious crime dropped 13%, part of a downward trend that began in the mid-1990s. He tried to ease minority tensions, which were high after the 1997 torture of Abner Louima and the deadly police shootings of Amadou Diallo in 1999 and Patrick Dorismond in 2000. Kerik accomplished this by requiring officers to attend community meetings, which he often also attended; reportedly, he received praise from local civil rights leaders (Hays, 2001).

September 11, 2001, Terrorist Attacks

The terrorist attacks on 9/11, provided the most formidable challenge in Kerik's career as NYPD Commissioner. The World Trade Center towers' collapse formed the infamous Ground Zero, where emergency workers (including police) sifted through debris for human remains and evidence for what was deemed the most extensive crime scene in history (Kerik, 2001).

As soon as Kerik heard about the attack on the morning of September 11, he drove to the scene. Initially, before the towers fell, he was on Liberty and Barclay Streets as people ran by screaming hysterically, and he witnessed people jumping from the burning towers. In the hours after the attack, Mayor Rudolph Giuliani, Kerik, and their aides narrowly escaped death when they were trapped inside a building at 75 Barclay Street near Ground Zero. At the time, Giuliani was on the phone with the White House. The mayor, Kerik, and their aides dodged debris falling from the towers. They were at this location because the city command center was located in 7 World Trade Center. The command center was also damaged by the falling debris and the collapse of the two main towers and collapsed later that day. The mayor, Kerik, and their aides made their way to Church Street and their cars to regroup. Kerik told the press his initial reaction was anger at the attacks (Marzulli, 2001a). He also said, "If anything happens to any one of us, there's a line of succession. Leaders should lead by example" (Marzulli, 2001a, p. 22).

According to media reports, the NYPD had the most sophisticated anti-terrorist strategies of any police force in the nation. They had participated in mock disasters, set up a high-tech command center, and advised the mayor on constructing a $7 million emergency bunker. However, the 9/11 attacks proved too overwhelming for the department's anti-terrorist preparedness. After the towers fell, phones and e-mail went dead at the command center at Police Plaza. The department was credited with innovating a response. Within hours of the disaster, Kerik called all off-duty officers to work. By day's end, shifts were elongated from 8 to 12 hours. Kerik also deployed an additional 10,000 officers—from narcotics officers to precinct detectives—to aid the 1,000 officers already at Ground Zero. Other officers continued patrolling New York City's streets; crime dropped 30% during the first few days of the crisis (Rashbaum, 2001). However, according to Kerik, bias attacks against people perceived as Arab increased in the 10 days after the attacks: a period during which there were 56 anti-Arab incidents, ranging from anti-Muslim graffiti to assaults on Muslims (Messing & Topousis, 2001).

Three days after 9/11, Kerik addressed the NYPD, expressing "the pride and gratitude I feel for what you have done over the past two days" (Marzulli & Standora, 2001, p. 45). At that point, 23 police officers were still missing (Marzulli & Standora, 2001); they have since been listed among the dead (Kerik, 2001). Kerik delivered an uplifting message, honoring the heroes of Ground Zero and expressing consolation to the victims' families (Marzulli & Standora, 2001).

The National Commission on Terrorist Attacks Upon the United States (also known as the 9–11 Commission) investigated New York's preparedness in their inquiry into the events of 9/11. On the whole, the 9–11 Commission found that the response was effective in saving the lives of those below the impact zone, due to the successful efforts on the day:

> Some specific rescues are quantifiable, such as an FDNY company's rescue of civilians trapped on the 22nd floor of the North Tower, or the success of FDNY, PAPD, and NYPD personnel in carrying nonambulatory civilians out of both the North and South Towers. In other instances, intangibles combined to reduce what could have been a much higher death total. It is impossible to measure how many more civilians who descended to the ground floors would have died but for the NYPD and PAPD personnel directing them—vla safe exit routes that avoided jumpers and debris—to leave the complex urgently but calmly. It is impossible to measure how many more civilians would have died but for the determination of many members of the FDNY, PAPD, and NYPD to continue assisting civilians after the South Tower collapsed. ("National Commission," 2004, p. 316)

The NYPD was not exempt from criticism, however. At hearings in New York, and in a draft report issued before the hearings, the panel criticized New York's emergency responders for failures and gaps in communication and the lack of coordination between the police and fire departments as well as their failure to convey information to 911 operators (a system supervised by the NYPD; Shenon & Flynn, 2004). For example, police helicopters circling the buildings did not transmit information to fire personnel in the buildings (Shenon & Flynn, 2004). The 911 operators and the fire department dispatchers also did not have enough information about the impact zone to properly advise callers trapped in the buildings ("National Commission," 2004). However, overall the 9–11 Commission praised NYPD's preparedness, efficient internal communication, and control of the scene, particularly in evacuating civilians who had descended from the towers and ensuring emergency vehicles' access to the site; the NYPD was also praised for having effective communications (radio) technology ("National Commission," 2004).

In another example of effective leadership, in the wake of the anthrax attacks in the fall of 2001, Kerik appointed Phil Pulaski, an NYPD investigator and chemical engineer, as the city's bioterrorism czar. Pulaski handled the anthrax threat at NBC headquarters (Weiss, 2001a).

Additionally, after the terrorist attacks, Kerik ordered mandatory mental health counseling for all 55,000 NYPD members, the first time this measure was ever required. By mandating it, Kerik removed the stigma associated with counseling, stating that voluntary counseling is often unutilized by officers for fear they would be perceived as weak. The program began on November 18, 2001, and required all NYPD staff to participate in group counseling. Second, an emergency telephone line and referral system provided additional anonymous counseling. Third, discussion groups convened and included spouses and family members affected by the attacks ("NYPD," 2001).

The NYPD also assisted the Justice Department in interviewing 86 Middle Eastern immigrants who allegedly had information regarding the 9/11 attacks. Kerik indicated that he would not ask about their immigration status, leaving that to the discretion of federal authorities ("NYPD to interview," 2001). In December 2001, Kerik testified in front of a Senate panel urging federal agencies to increase information sharing with local police departments. He advocated a revision of the security clearance system, which granted clearance to some cops while denying it to the cops' superiors (Port, 2001). He stated, "Continuing to maintain walls between federal and state authorities . . . represents the worst kind of dysfunctional thinking in government" (Port, 2001, p. 6).

Ironically, Kerik completed his memoirs (for publication) the night before the attacks; however, he added an epilogue describing his role in the response to the attacks in the weeks that followed. The book, *The Lost Son* (2001), was rushed to the press for release in November 2001 in order to maximize sales. Some, including City Councilman Sheldon Leffler, believed Kerik was trying to profit from a tragedy (Hays, 2001). A *Newsweek* columnist called the book "unconscionable," and another stated it was a "conflict of interest and an abdication of his responsibilities as police commissioner" (Pyle, 2001). However, according to the media, Kerik, Mayor Giuliani and Fire Commissioner Thomas Van Essen, had become national symbols of heroism after the attacks (Hays, 2001). At a news conference about 9/11 on November 8, 2001, Kerik stated,

I shouldn't get credit for what happened that day. The everyday cops—they're the ones who should get the credit. . . . This department will get along without me. (Hays, 2001)

Kerik defended the publication of his book, arguing that it was a memoir. He was police commissioner during 9/11, a historical event in the nation's history; therefore, it must be part of his story. On November 10, 2001, he told the press,

I could say at least three times, between Mayor Giuliani and myself, we came close to not making it out of there. So I think it would be remiss not to write it, or say something. (Pyle, 2001)

He also spoke on the *Oprah Winfrey Show:*

Being the police commissioner after September 11 is painful. It's heart-wrenching. But it also has clearly demonstrated to me that this is the best job you could ever have. Because the people that work for [me], there are none better. ("Horror and heroes," 2001)

However, another scandal was associated with the book. From May to October 2001, Kerik used three officers to help him discover how his mother died in Ohio in 1964. The use of that manpower violated an agreement he had made with New York City's conflicts of interest board, which allowed him to write his book but which forbade him from using city time or his official position to obtain any private or personal advantage for himself or the publisher (De la Cruz, 2002). Kerik was fined $2,500 for using the officers to do research for his book.

Kerik announced that he would step down from his position as commissioner on January 1, 2002, despite mayoral candidates Michael Bloomberg, Alan Hevesi, Herman Badillo, and Peter Vallone indicating that they wanted to keep him in the position (Colangelo, 2001).[1] Subsequently, Mayor-Elect Bloomberg publicly indicated that he wanted Kerik to remain as commissioner (Weiss, 2001b). But Kerik turned down the offer in order to spend more time with his family (Hays, 2001). In December 2001, Kerik promoted ten captains, four deputy inspectors, and one full inspector, rewarding leaders associated with drops in crime rates in their respective precincts (Messing, 2001).

In June 2002, Kerik became chairman of the New York State Athletic Commission. His responsibilities included overseeing boxing and wrestling matches, issuing licenses, enforcing rules and regulations, and reviewing medical and safety standards ("Ex-NYC police chief," 2002). In 2003, he was the interim Minister of the Interior and senior policy advisor in Iraq after its liberation by U.S. forces. He rebuilt police and emergency services in that country ("Bernard Kerik," n.d.).

Kerik's ability to respond competently to the 9/11 attacks was clearly related to experiences of his early childhood. He was no novice to traumatic situations, and his early-childhood family experiences shaped his leadership qualities and responses. According to Northouse (2004), childhood family experiences create leaders who always take a paternal role with subordinates. Based on family experiences, leaders exhibit behaviors along continuums of authoritative to permissive, supportive to critical, and dependent to counterdependent to independent. Kerik's response in the aftermath of the 9/11 attack represents a combination of the continuums.

Kerik was authoritative when he called all off-duty officers to work, extended shifts, and deployed 10,000 officers to support the efforts at Ground Zero; he was supportive when he ordered mandatory mental health counseling, anonymous forms of support, and discussion groups for spouses and family members. He also showed independent decision making by taking an unorthodox position on immigration issues when suspects were questioned at a time when such behavior might have been perceived as antagonistic.

Another example shows Kerik's leadership: his support for the Police Studies Certificate (cocreated and cocoordinated by the author), which gave NYPD police officers access to courses teaching supervisory leadership skills in a multicultural environment and was the best testimony of his relentlessness of commitment to his troops. The program started the week before 9/11. The first two weeks after the 9/11 events, officers were absent from classes. Yet by week three, they were returning, to continue classes, and to enhance their skills to serve the public better—a tribute to their colleagues who died. Commissioner Kerik encouraged the continuation of the program despite the other demands pulling at officers during the troubled weeks and months immediately following 9/11. By the end of the semester, most of the officers had completed the courses.

Although it is unlikely that he was consciously aware of it, Kerik used a formula that can be best summarized by Clawson's (2003) six steps to effective leadership.

His life mission was to become an innovative leader, and his career path points to his various attempts to become successful in many fields. After clarifying what was possible in a given environment and what others could contribute to his success, Kerik assigned them roles in which they could be successful. He was ready to give credit to and to support his subordinates. For example, Kerik continued to support the Police Studies Certificate program and the enrolled officers despite the temptation to cut extracurricular activities for officers who had just experienced the worst disaster of their careers. Finally, prior to his departure, he made sure that the officers who contributed to his success were properly acknowledged and promoted.

According to the psychodynamic approach to leadership, it takes someone who has experienced much hardship in life to deliver the authoritative yet supportive qualities depicted by Kerik. As Clawson (2003) posits, "Behaving as and becoming

Six Steps to Effective Leadership

Step 1: Clarifying your center
What is your life's mission?

Step 2: Clarifying what's possible
What can really be achieved?

Step 3: Clarifying what others can contribute
What can others really do?

Step 4: Supporting others so they can contribute
What resources can you really commit?

Step 5: Being relentless
What degree of confidence and commitment do you/others have?

Step 6: Measuring and celebrating progress
How do you praise, and what kind of positive feedback do you provide?

(Clawson, 2003).

an effective leader is a secondary by-product of an intense commitment to a purpose" (p. 94). It is hard to question the intense commitment Kerik exhibited to making the NYPD the best it could be.

In November, 2007, Kerik plead not guilty to numerous federal charges including corruption, conspiracy, and tax fraud, and posted $500,000 bail (Buettner & Stowe, 2009; Weiser, 2008). Federal District Court Judge Stephen C. Robinson revoked his bail in October 2009 because he provided sealed information to a New Jersey lawyer (Buettner & Stowe, 2009).

In a plea deal with prosecutors, Kerik plead guilty to eight felonies including: tax fraud, lying to federal investigators, lying on a loan application and a questionnaire for a government position, hiding profits and royalties from his book "The Lost Son," and lying about $255,000 of work done on his home in exchange for the prosecution dropping the major charges against him and redacting its request that he face three separate federal trials (Barish, 2009; Moritz, 2010).

On February 18, 2010, Kerik was sentenced to four years in federal prison and ordered to pay $188,000 in restitution and to pay past due taxes and penalties for the preceding six years ("Bernard," 2010; "Feds," 2010). He appealed his sentence to the 2nd U.S. Circuit Court of Appeals which affirmed on March 31, 2011 (Neumeister, 2011).

Kerik is currently serving a four-year prison sentence in a federal prison in Cumberland, MD, and will be released in October 2013 (Golding, 2011; Moritz, 2010).

[1]Two other candidates, Mark Green and Fernando Ferrer, who were often critical of the NYPD under Kerik's command did not favor his continuing as commissioner (Pyle, 2001).

Exercise

The police commissioner detailed in this chapter has been examined through the prism of the psychodynamic approach. Using what you have learned from other theories in this book or from the summaries (as indicated in the Introduction) you have read thus far, it is possible to consider his approach through the lens of different theories. What other theories can you find applicable to this chief? Are there lessons to be learned from those applications, or do those theories indicate alternative paths or approaches that might have yielded a more successful outcome?

References

Adler, A. (1928). Characteristics of the first, second and third child. *Children*, 3, 14–52.

Andeweg, R.B. & S.B. Van Den Berg (2003). Linking birth order to political leadership: The impact of parents or sibling interaction? *Political Psychology*, 24 (3), 605–623.

Barish, S. (2009, November 5). Kerik pleads guilty to lying to feds, admits tax crimes. *Los Angeles Times*. Retrieved May 25, 2011, from http://www.latimes.com/news/nation world/nation/wire/wpix-bernard-kerik-pleads-guilty,0,130377.story.

Bass, B.M. (1990). *Bass and Stogdill's handbook of leadership: A survey of theory and research*. New York: Free Press.

Bellin, G. & M.W. Rennie (2004). People want to develop. *McKinsey Quarterly*, 2, 37–39.

Bernard Kerik (n.d.). *Washington Speakers Bureau*. Retrieved January 23, 2003, from www.washspkrs.com

Bernard Kerik sentenced to four years in prison. (2010, February 18). *The Washington Post*. Retrieved May 25, 2011, from http://voices.washingtonpost.com/44/2010/02/bernard-kerik-sentenced-to-fou.html.

Berne, E. (1961). *Transactional analysis in psychotherapy*. New York: Grove.

Buettner, R. & S. Stowe (2009, October 20). Ex-commissioner Kerik is jailed as judge assails pretrial conduct. *The New York Times*. Retrieved May 25, 2011, from http://www.nytimes.com/2009/10/21/nyregion/21kerikhtml.

Clawson, J.A. (2003). *Level three leadership: Getting below the surface* (2nd ed.). Upper Saddle River, NJ: Prentice Hall.

Colangelo, L. (2001, October 5). It's goodbye to NYPD on Jan. 1. *Daily News*, p. 39.

De la Cruz, D. (2002, February 28). Kerik agrees to pay fine after conflict of interest found. *The Associated Press, state and local wire.*

Ex-NYC police chief picked to run group. (2002, June 21). *The Times Union*, p. C2.

Feds want former NY top cop to go directly to jail. (2010, February 8). *The Seattle Times.* Retrieved May 25, 2011, from http://seat tletimes.nwsource.com/html/nation world/2011018977_ apuskerikinvestigation.html.

Former police commissioner finds long-lost daughter. (2002, April 11). *The Associated Press*, state and local wire.

Freud, S. (1921; 1955). *Group psychology and the analysis of the ego.* London: Hogarth Press.

_____(1939; 1964). *Moses and monotheism.* London: Hogarth Press.

George, A. & J. George (1998). *Presidential personality and performance.* Boulder, CO: Westview Press.

Golding, B. (2011, April 1). Judges rip appeal bid by Kerik. *New York Post.* Retrieved May 25, 2011, from http:// www.nypost.com/p/news/local/manhattan/judges_rip_ appeal_bid_by_kerik_OoFwG7n6obVuM6qxMK4EuJ.

Hays, T. (2001, November 9). Kerik announces he'll leave NYPD. *The Associated Press*, state and local wire.

Hogan, R., G.J. Curphy & J. Hogan (1994). What we know about leadership: Effectiveness and personality. *American Psychologist*, 49 (6), 493–504.

Horror and heroes: What he saw at Ground Zero. (2001). *The Oprah Winfrey Show.* Retrieved January 23, 2004, from www.oprah.com.

Kerik, B. (2001). *The lost son: A life in pursuit of justice.* New York: Harper-Collins.

Kets de Vries, M.F.R. (1980). *Organizational paradoxes: Clinical approaches to management.* New York: Tavistock.

Kohut, H. (1976). Creativeness, charisma, group psychology: Reflections on the self-analysis of Freud. In P.H. Ornstein (Ed.), *The search for the self* (pp. 793–843). New York: International Universities Press.

Kroeger, O. & J.M. Theusen (1992). *Type talk at work.* New York: Delacorte.

LaBarre, W. (1972). *The ghost dance: Origins of religion.* New York: Delta.

Mann, R.D. (1959). Review of the relationship between personality and performance in small groups. *Psychological Bulletin*, 56 (4), 241–270.

Marzulli, J. (2001a, September 27). Giuliani & group of top cops barely dodged death as towers fell. *Daily News*, p. 22.

_____(2001b, October 30). Kerik book with WTC story is rushed. *Daily News*, p. 14.

_____ & L. Standora (2001, September 15). Kerik tells troops of NYPD's great tragedy: Says the list of officers missing grows to 23. *Daily News*, p. 45.

Messing, P. & T. Topousis (2001, September 21). Anti-Arab assaults surge here. *The New York Post*, p. 22.

Moritz, O. (2010, May 18). Disgraced ex-top cop Bernard Kerik becomes inmate number 84888-054 in Maryland federal prison. *Daily News.* Retrieved May 25, 2011, from http://articles.nydailynews.com/2010-05-18/news/ 27064653_1_inmate-number-top-cop-bernard-kerik.

National Commission on Terrorist Attacks Upon the United States. (2004). *The 9/11 commission report.* Washington, D.C.: U.S. Government Printing Office. Retrieved August 10, 2004, from http://www.9–11com mission.gov/report/index.htm.

Neumeister, L. (2011, March 31). Bernie Kerik's four-year prison sentence upheld. *Huffington Post.* Retrieved May 25, 2011, from http://www.huffingtonpost.com/ 2011/03/31/bernie-keriks-4year-priso_n_843155.html.

Northouse, P.G. (2004). *Leadership: Theory and practice* (3rd ed.). Thousand Oaks, CA: Sage.

NYPD orders entire department to undergo mandatory counseling. (2001, November 30). *The Associated Press*, state and local wire.

NYPD to interview 86 immigrants on Justice Department's list. (2001, December 5). *The Associated Press*, state and local wire.

Port, B. (2001, December 12). Kerik to pols: Let FBI, cops share info. *Daily News*, p. 6.

Pyle, R. (2001, November 11). Police commissioner brush-es off critics of book. *The Associated Press*, state and local wire.

Rashbaum, W.K. (2001, September 17). After the attacks: Safety. *The New York Times*, p. A7.

Shenon, P. & K. Flynn (2004, May 18). Panel has a question: Why wasn't the city prepared? *The New York Times*, p. B1.

Stech, E.L. (2004). Psychodynamic approach. In P.G. Northouse (Ed.), *Leadership: Theory and practice* (3rd ed.). Thousand Oaks, CA: Sage.

Strozier, C. & D. Offer (1985). Introduction. In C.B. Strozier & D. Offer (Eds.), *The leader: psychohistorical essays.* New York: Pergamon.

Weber, M. (1946; 1973). Politics as a vocation. In *From Max Weber: Essays in sociology* (H.H. Gerth & C. Wright Mills, Trans.). Oxford: Oxford University Press.

_____(1947; 1964). *The theory of social and economic orga-nization* (A.M. Henderson & T. Parson, Trans.). New York: Oxford University Press.

Weiser, B. (2008, December 29). Kerik pleads not guilty to new charges. *The New York Times.* Retrieved May 25,

2011, from http://cityroom.blogs.nytimes.com/2008/12/29/kerik-pleads-not-guilty-to-new-charges/.

Weiss, M. (2001a, October 19). Kerik taps czar for bioterror. *The New York Post,* p. 22.

_____ (2001b, November 1). Bloomy wants Kerik in his corner. *The New York Post,* p. 9.

Whyte, D. (1994). *The heart aroused: Poetry and the preservation of the soul in corporate America.* New York: Doubleday.

Winer, J.A., T. Jobe & C. Ferrono (1984–1985). Toward a psychoanalytic theory of the charismatic relationship. *Annual of Psychoanalysis,* 12–13, 15–175.

Zaleznik, A. (1974). Charismatic and consensus leaders: A psychological comparison. *Bulletin of the Menninger Clinic,* 38 (3), 222–238.

_____(1977, May–June). Managers and leaders: Are they different? *Harvard Business Review,* 55 (3), 67–68. (Reprinted in 1992, *Harvard Business Review,* 70 (2), 126–135.)

Soliciting and Entertaining 100 Ideas
Skills Approach

OVERVIEW OF THE APPROACH

The skills approach focuses on the individual leader and centers on the development of leadership skills. Due to the many innovations he brought to the Boston Police Department, Commissioner Paul F. Evans was chosen to illustrate this approach.

Skills are learned and developed over time and are a function of both education and experience (Mumford et al., 2000b). They are capacities that are beyond mere intelligence. Here cognitive abilities such as the exceptional capacity at written and oral expression, comprehension, and the ability to find and manipulate information are more important than intelligence (Bass, 1990). Leadership is the one combination of knowledge, problem-solving skills, solution construction skills, and social judgment needed to solve problems (Mumford, Zaccaro, Connelly, & Marks, 2000a).

LEARNED LEADERSHIP SKILLS

The notion that leadership skills can be learned was first articulated by Sheriff (1968), who challenged the dominant leadership mythology that emphasized trait theory (see Chapter 12). According to Sheriff (1968), possessing intelligence is a prerequisite of leadership; however, all other leadership qualities can be learned via skill development. There are three types of skills: technical, human, and conceptual. Technical skills are methods and techniques associated with a profession. Human skills reflect a capacity to work effectively in an organization. Conceptual skills imply the ability to generate, consider, and use ideas. Sheriff (1968) also posited that communication abilities, broad and well-rounded interests, problem-solving abilities, mental and emotional maturity, and motivation are associated with leadership. Sheriff (1968) states

> What is significant is that leadership is not simply a matter of innate personality or physical traits. It appears more profitable to view leadership in terms of a constellation of acquired personality traits and developable skills (p. 35).

Since skills are developable, training programs help grow leaders. Pernick (2001) advocates organizational training programs for subordinates who could become leaders. The advantages of this approach are that "the organization gets to groom the next generation in line with its culture and strategic agenda . . . [and that] the organization has a greater control over the supply of leaders with the requisite skills, making strategic implementation faster" (Pernick, 2001, p. 11).

LEADERSHIP IN CONTEXT

The skills approach places leaders' development in the context of the demands of their jobs. Analysis of leader's skills begins with identifying the demands placed on them (Fleishman & Quaintance, 1984). Leadership problems are non-routine and are characterized as novel, complex, and ill-defined. It is here that leadership is most crucial. Here, "unless leaders can identify significant organizational problems and formulate solutions to these problems, all the planning and all the persuasion in the world are to no avail" (Mumford et al., 2000a, p. 157). Likewise, a leader's capacity to handle information when solving problems is primary:

> There is a plethora of available information in complex organizational systems, only some of which is relevant to the problem. Further, it may be difficult to obtain accurate, timely information and identify key diagnostic information. As a result, leaders must actively seek and carefully evaluate information bearing on potential problems and goal attainment. (Mumford et al., 2000b, p. 14)

Leadership and Intelligence

Intellectual intelligence (IQ)
- It is genetic.
- It is revealed in curiosity.
- It is honed by discipline.
- It is supported by a range of experiences.

Emotional intelligence (EQ)
- Recognize your own emotions.
- Manage your own emotions.
- Control yourself.

Social intelligence (SQ)
- Recognize emotions in others.
- Listen.
- Care about others' emotional states.
- Help others gain control and manage their emotions.

Change intelligence (CQ)
- Recognize the need for change.
- Understand the change process.
- Master the change process.
- Find comfort in managing the change process.

(Clawson, 2003)

Leaders operate in real-world settings were the conditions for problem solving are seldom ideal and often restricted time wise. Leaders must also be aware of the potential consequences of their decision(s) and all relevant stakeholders' (Mumford et al., 2000b). In essence, they must be conscious of the social context of their problem-solving endeavor; this places social skills at the center of the approach (Zaccaro, Gilbert, Thor, & Mumford, 1991). Effective leadership, therefore, is dependent on the leaders' ability to solve social problems in organizations despite organizational constraints (Yammarino, 2000).

This approach also involves the willingness of leaders to place themselves in situations in which they can learn, making skill development a function of leaders' interaction with their environment. Certain personality characteristics, such as dominance, may be associated with seeking out these situations (Lord, Devader, & Alliger, 1986). Mumford et al. (2000b) identified four types of experiences that are most conducive to leadership development: 1) job assignments with novel, challenging problems; 2) mentoring of subordinates; 3) leadership training; and 4) hands-on experience in solving organizational problems.

Since experience is central to skill development, skills are slow to develop. According to Ericsson and Charness (1994), it takes approximately 7 to 10 years to acquire the skills needed to solve leadership problems; however, a later study by Mumford, Marks, Connelly, Zaccaro, & Reiter-Palmon (2000c) suggests that it may take 20 years to acquire these skills. Additionally, the types of problems encountered by senior management are more beneficial in developing leadership skills. Also higher levels of problem solving and social judgment associated with leadership are required at top levels (Mumford et al., 2000b). Therefore, leadership skills are built up through the problem-solving experiences of leadership itself.

SOCIAL JUDGMENT AS SOCIAL INTELLIGENCE

A component of leadership capacity is social judgment, which is similar to social intelligence. Social intelligence was first identified by E. L. Thorndike in 1920 as the ability to act wisely in human relations. Zaccaro et al. (1991) elaborated on this, proposing that social perceptiveness and behavior flexibility are the attributes of effective leadership. Social intelligence is distinct from general intelligence, and social perceptiveness refers to leaders' sensitivity to the plights of both subordinates and the organization. Current researchers define social intelligence as 1) the ability to perceive others' needs and problems and 2) the ability to respond and adapt to different environments (Kobe, Reiter-Palmon, & Rickers, 2001). For them, behavior flexibility is defined as a leader's capacity to respond appropriately and flexibly based on the situation (Kobe et al., 2001).

Kobe et al. (2001) studied 192 undergraduates in order to determine whether people with high social intelligence were more likely to engage in leadership activities. They also tested whether emotional intelligence (see below), as a subset of social intelligence, contributed to leadership participation above and beyond that of mere social intelligence. What they found was that social intelligence was correlated with leadership participation, which suggests that it may be a primary component of leadership ability and effectiveness. However, emotional intelligence added to the variance in leadership participation above and beyond social intelligence.

EMOTIONAL INTELLIGENCE

Kobe et al. (2001) characterized emotional intelligence as a type of social intelligence. Emotional intelligence is measured using Bar-On's Emotional Quotient Inventory (EQ-i). Feldman (1999) characterized emotional intelligence as the ability to apply emotional and social understanding to positively influence subordinates. Feldman (1999) identified core skills and higher-order skills associated with emotional intelligence; the former included: knowing yourself, maintaining control, being able to read others, possessing accurate perception, and communicating with flexibility; the latter included taking responsibility, generating options and alternatives, formulating a vision, possessing courage, and having resolve.

Emotional intelligence is associated with the work of Goleman (1998), who defined it as the capacity to be aware of and to handle emotions in various situations. Goleman (1998) posited that emotional intelligence is a central contributor to effective leadership; likewise, one's emotional intelligence quotient (EQ) can be raised via the learning process. The learning process is specific to the context of the leader, which involves learning about group interaction and information sharing, and describes the potential for emotions to be contagious within organizations. EQ is increased via behavior change. For Goleman (1998), the positive feedback that results from understanding and controlling emotions in the workplace positively reinforces the change (Dearborn, 2002).

U.S. ARMY STUDIES

In a study of approximately 1,800 U.S. Army officers, complex problem-solving skills, social judgment skills, and leader knowledge were found to predict the achievement of leaders in solving complex problems. Using the same sample, another study concluded that knowledge, problem-solving skills, and social skills increased as experience increased. Assignments, training, and career development all contributed to the bettering of leaders' skills over time. Specific experience was more beneficial at particular points of leaders' careers, suggesting that the path to leadership success is complex (Mumford et al., 2000c).

SKILLS APPROACH TO LEADERSHIP AND THE CASE OF COMMISSIONER PAUL F. EVANS

The notion of skills approach to leadership is directly associated with the concept of human intelligence as a function of learning and development. Most of us grew up socialized to the concept that people were either smart or stupid (or somewhere in the middle); on a more sophisticated level, people who were smart or stupid were referred to as having a high or low level of intelligence. Here, intelligence was perceived as being similar to physical qualities such as eye color and hair color; likewise, it was assumed that you were doomed to live with the amount of intelligence with which you were born.

Today, social scientists look at intelligence as a social phenomenon; they have identified four categories of intelligence: intellectual, emotional, social, and change.

Leadership style can be viewed through this new approach to the concept of human intelligence. Intellectual intelligence (IQ) is revealed in curiosity and is honed by discipline and supported by a range of experiences. Emotional intelligence (EQ) depends on the ability to recognize and handle one's own emotions and the level of control one exercises over those emotions. Social intelligence (SQ) is the ability to recognize emotions in others, to listen and care about others' emotional states, and to help others to gain control and manage their emotions. Finally, change intelligence (CQ) is the ability to recognize the need for change, to understand the change process, and the level of comfort in managing it (Clawson, 2003). A successful leader (SL) combines these four elements, as shown in the following formula:

$$SL = High\ IQ + High\ EQ + High\ SQ + High\ CQ$$

This is not to say that anyone who does not fit the above formula cannot be a good or successful leader, but the degree to which someone is deficient in one of the variables will significantly affect his/her overall leadership success.

Commissioner Paul F. Evans illustrates the above formula. Choosing Commissioner Evans for this chapter was partially motivated by an objective evaluation of his career and by a minimal yet influential encounter between the author and Commissioner Evans at a closed conference in 1999, which allowed the author to draw conclusion as to his leadership style.

CAREER HIGHLIGHTS

Skills Approach

Commissioner Paul F. Evans
Boston Police Department, Massachusetts

Paul F. Evans grew up in South Boston (Cullen, 1996; Negri, 1994). From 1967 to 1969, he was a corporal in the U.S. Marines and served in the Vietnam War. After returning from the war, he went to college and received a bachelor's degree in political science, in 1974, from Boston State College. In 1978, he earned his doctor of law degree (cum laude) from Suffolk University Law School ("Paul F. Evans," n.d.).

Evans joined the Boston Police Department (BPD) in 1970. From 1986 to 1992, he served as the superintendent in chief of field services. Subsequently he served as superintendent in chief of the Bureau of Investigative Services. From July 1993 to February 1994, Evans was second in command under Commissioner William Bratton, serving as superintendent in chief ("Paul F. Evans," n.d.).

In February 1994, Evans was named acting commissioner when Bratton left the position; he served in that position until November 2003 (Belkin, 2003; DeMarco, 2003; Lakshmanan, 1994). Evans called a multilevel agency meeting early in his tenure in order to coordinate crime-fighting efforts and promote his vision of neighborhood (community-oriented) policing (Lakshmanan, 1994). There, he outlined his belief that neighborhood policing was a restoration of a style of policing prevalent when he was growing up in South

Boston in the 1950s and 1960s. This vision included making uniformed officers more visible and building their integrity in order to make them role models for youths ("Evans takes charge," 1994). He also advocated that the same cops work the same neighborhoods, so they could build trust with local residents (Mallia, 1994).

In pursuit of his neighborhood-policing vision, Evans decentralized the citywide drug unit ("A new openness," 1995). He also increased the number of neighborhood police districts from 5 to 10, actively calling on private citizens to help set goals for the department ("A new openness," 1995; Cullen, 1996). Evans also hired 300 additional officers, acting in a neighborhood capacity, during his first two years as commissioner (Cullen, 1996). In June 1995, Evans released a list of 100 ideas about public safety compiled from 16 teams of community leaders, including clergy, politicians, and police officers. The goals included developing strategies to deal with delinquent youths and redesigning police stations to make them more appealing for area residents seeking help (Valliancourt, 1995).

Initiatives to curb youth violence were a priority throughout Evans's tenure as commissioner. He attended President Clinton's unveiling of the federal initiative to computerize records and track guns confiscated in juvenile criminal investigations.

This initiative was modeled after Evans's program, Operation Scrap Iron, in which gun suppliers and dealers, as well as young offenders, were targeted (McGrory, 1996). In 1995, BPD teamed up with local YMCAs to identify 300 at-risk youths, who were then sent to summer camps. Between 1993 and 1997, Boston was nationally known for stemming the tide of crime by youths (ages 14 to 24) by 9% (Menendez, 1998).

Evans often spoke in press conferences about the disconnect between crime statistics and citizens' perception of their own safety. Although overall crime was dropping in the late 1990s, Evans indicated that the simultaneous spike in juvenile violence may have negated any sense of safety citizens felt. "Perception is reality. If people don't feel safe, it doesn't matter what the stats say," he said (Cullen, 1996, p. 14). In 2002, Evans told reporters that despite high-profile incidents, such as the murder of 10-year-old Trina Persad on June 2 and a series of sexual assaults, violent crime was at a statistical 31-year low. However, he also recognized that high-profile incidents serve to produce citizen fear (Ranalli, 2002; Rosenwald, 2002a). Evans stated:

You have the statistics and you have the perception. But perception is reality. I'm not going to downplay any of the high-profile cases we've had in this city. . . . But people's perceptions come from what they see on television. What we have to do is reduce fear, and we're committed to doing that, and solving all these cases. (Rosenwald, 2002a, p. A1)

As commissioner, Evans received local praise from the mayor, politicians, and the press for his leadership during two crises. In a March 1994 incident, a botched police drug raid led to the death of a Dorchester neighborhood minister when officers mistakenly entered his apartment instead of another dwelling, startling him and causing a heart attack. Evans apologized the next day, an unusual move considering the potential legal liability. The apology was praised by members of the black community

(Cullen, 1996). Evans subsequently suspended the responsible officer for 15 days for improperly relying on an informant who had given him the wrong address (Canellos, 1995).[1] In May 1994, an officer was arrested for raping a prostitute, which prompted an immediate review of BPD policies that addressed police misconduct (Estes, 1994). The officer was immediately put on leave (Chacon, 1996).

Evans fared less well in another high-profile incident on January 25, 1995, in which a black undercover cop (on duty), Michael Cox, was beaten by fellow cops who mistook him for a suspect. Cox was in plainclothes chasing a suspect when he was struck from behind by five uniformed cops who thought he was the target. According to court records, once the officers realized they had beaten a cop, they fled the scene ("Judge finds," 1998). Cox suffered a concussion, multiple face cuts, and kidney damage and could not work for six months (Noonan, 1999). A federal judge ruled that the incident constituted a civil rights violation. Evans ultimately conceded that Cox's civil rights were indeed violated; however, his (and the city's) lawyer indicated that this did not mean there was a pattern of brutality in BPD ("Judge finds," 1998). The city ultimately settled with Cox for $700,000 ("Evans must act," 1999). The incident and subsequent lawsuits led to criticism that there was a pattern of racist conduct in BPD, despite Evans's declaration that thereafter such conduct would not be tolerated. In another incident in 1999, two white officers racially harassed a black officer, hanging a noose made of yellow police tape over his motorcycle ("Evans must act," 1999).

These racial incidents sparked black community leaders to call for a civilian review board. Evans was against the creation of such a board, arguing that complaints should be handled within the department. He also said, "Racism and harassment are totally alien to the oath we swear to and won't be tolerated" ("Black leaders," 1999).

Evans also was confronted with a lawsuit initiated by eight white cops who claimed they were overlooked for promotions in favor of the promotions of three black officers, based on race.

Evans argued that race-based promotions were necessary in order to ensure a racially diverse police force that reflected the citizens of Boston (Martinez, 2002).

Despite much praise by people outside BPD, Evans has been considerably criticized from within. He has largely been seen as a commissioner who bows to public opinion (Rosenwald, 2002b). His relationship with BPD officers first soured in 1995 when he invoked a city provision to disregard union contracts in assignment and deployment in order to accomplish the decentralization of the citywide drug unit (Ford, 1995; Heaney, 1997; "Stand with Evans," 1998). In 1997, his relationship with patrol officers reached its nadir when a bill went before the city to prevent the commissioner from being able to circumvent union contracts in the name of public safety. Evans, along with the Massachusetts Chiefs of Police Association, opposed the bill (Heaney, 1997). Legislators ultimately sided with the patrol officers, stripping Evans of the power to override collective bargaining agreements ("Controlling the police," 2002).

In 2002, the Boston Police Patrolman's Association voted "no confidence" in Commissioner Evans after he banned BPD officers from shooting at moving cars unless passengers shot at cops first ("Hub patrolmen," 2002). The policy change was prompted by the death of a passenger in a fleeing car who was shot and killed by a police officer on September 8, 2002. By late September 2002, the patrol officers' union was calling for Evans's resignation. They indicated that the ban on shooting at fleeing automobiles would put their lives at risk. However, community leaders, including the Black Ministerial Alliance, supported Evans ("More support," 2002). Additional supporters were the Garrison Trotter Neighborhood Association, the Boston Chamber of Commerce, Suffolk County District Attorney Dan Conley, and two independent candidates for district attorney (Szaniszlo, 2002; Woodlief, 2002).

Evans, when questioned about his relationship with the rank and file, stated that his top allegiance was to the city. As *The Boston Globe* reported:

In a town of mythic police solidarity, [Evans] has sought to create a department accountable not only to itself, but to the broadest possible constituency. It's a mandate that has earned him uncommon respect outside the department's walls but also hardened dissent among the very ranks he has held, and climbed, over 32 years. . . . Today, many regard Evans as one of the most progressive police managers in the country. But to earn that regard, he has had to shut down the role that long defined being chief: defending and insulating officers, at any cost, from a critical public. (Rosenwald, 2002b, p. B1)

As was the case in many police departments around the country, the economic recession of the early 2000s meant that Evans had to make do with fewer officers due to budget constraints. In December 2002, he was forced to cancel a class of 60 police recruits because of a hiring freeze in anticipation of a $100 million cut in state aid (Marantz, 2002). In November 2003, Evans retired from the BPD to take a position advising the United Kingdom on police leadership practices (Belkin, 2003; DeMarco, 2003; Sennott, 2004). A nationwide search was initiated to find his successor, who was appointed in early 2004 (Rosenwald & Estes, 2004).

Evans was the commissioner of BPD for almost 10 years. It takes a very skillful individual to head a large city with very serious problems for a full decade and leave the position without accusations related to major scandals. Nothing illustrates his skills approach to leadership better than the following account, which distills the essence of all four variables from the successful leader formula. A year after his appointment, he released a list of 100 ideas about public safety in Boston, compiled from 16 different teams of community leaders. Such an initiative not only is unprecedented in the history of police leadership but also exemplifies the combination of Evans's ability to recognize the need for change and to recognize that the change can be achieved only by factoring in how others feel about it. Such recognition in turn requires the ability to control one's own emotions, feelings, and vision for a department. All these skills were definitely honed by Evans's military discipline and experience. How much of it was genetic? One can only guess. However, the right combination of all the variables was definitely present, making Commissioner Evans an example of the skills approach to leadership.

Between 2003 and 2007, Evans was the Director of the Police and Crime Standards Directorate of the Home Office in the United Kingdom ("Former," 2009; GardaWorld, 2008). During his tenure, he advised two Prime Ministers and four Home Secretaries on crime issues (GardaWorld, 2008). He returned to the United States, in 2007, and started his own international consulting practice before being appointed Managing Director of GardaWorld's Boston Office, which is part of Garda World Security Corporation ("Former," 2009; GardaWorld, 2008). He was later hired as a Public Safety Consultant by Penobscot Bay Media and as of 2009, was working as a consultant for Suffolk Downs, an East Boston racetrack that is

considering expanding its business ventures to include a resort casino ("Former," 2009; Technology Association of Maine, 2009).

[1]After this incident, a minority of politicians and community leaders criticized the ostensibly light punishment of the officer responsible. One senator, Dianne Wilkerson, told reporters that a civilian review board should be implemented in order to address the "growing divide between the black community and urban police departments" (Canellos, 1995, p. 14).

Exercise

The police commissioner detailed in this chapter has been examined through the prism of the skills approach. Using what you have learned from other theories in this book or from the summaries (as indicated in the Introduction) you have read thus far, it is possible to consider his approach through the lens of different theories. What other theories can you find applicable to this commissioner? Are there lessons to be learned from those applications, or do those theories indicate alternative paths or approaches that might have yielded a more successful outcome?

References

A new openness at the BPD. (1995, March 22). *The Boston Globe*, p. 12.

Bass, B.M. (1990). *Bass & Stogdill's handbook of leadership: Theory research and managerial application* (3rd ed.). New York: Free Press.

Belkin, D. (2003, November 13). Evans won't name choice for successor. *The Boston Globe*, B4.

Black leaders urge Boston Police Department review board, (1999, June 12). *The Associated Press*, state and local wire.

Canellos, P.S. (1995, October). Evans defends action against officer in raid, minister's lawyers criticize suspension. *The Boston Globe*, p. 14.

Chacon, R. (1996, February 11). Commissioner takes a new tack this time. *The Boston Globe*, p. 44.

Clawson, J.A. (2003). *Level three leadership: Getting below the surface* (2nd ed.). Upper Saddle River, NJ: Prentice Hall.

Controlling the police. (2002, September 14). *The Boston Globe*, p. A12.

Cullen, K. (1996, January 7). The neighborhood cop. *The Boston Globe*, p. 14.

Dearborn, K. (2002). Studies in emotional intelligence redefine our approach to leadership development. *Public Personnel Management*, 31 (4), 523–530.

DeMarco, P. (2003, October 26). A look back at the Evans years. *The Boston Globe, City Weekly*, p. 9.

Ericsson, K.A. & W. Charness (1994). Expert performance: Its structure and acquisition. *American Psychologist*, 49, 725–747.

Estes, A. (1994, May 2). Menino stands by police chief, department. *The Boston Herald*, p. 7.

Evans must act quickly. (1999, April 27). *The Boston Herald*, p. 26.

Evans takes charge. (1994, February 14). *The Boston Globe*, p. 22.

Feldman, D.A. (1999). *The handbook of emotionally intelligent leadership: Inspiring others to achieve results*. Falls Church, VA: Leadership Performance Solutions Press.

Fleishman, E.A. & M.K. Quaintance (1984). *Taxonomies of human performance: The description of human tasks*. Potomac, MD: Management Research Institute.

Ford, B. (1995, February 14). Highs & lows mark top cop's first year. *The Boston Herald*, p. 6.

Former Boston police commissioner Evans to advise Suffolk Downs. (2009, September 23). *The Boston Globe*. Retrieved May 26, 2011, from http://www.boston.com/news/local/breaking_news/2009/09/former_boston_p_1.html.

GardaWorld appoints former Boston police commissioner Paul Evans as managing director. (2008, October 6). *Reuters*. Retrieved May 26, 2011, from http://www.reuters.comarticle/2008/10/06idUS124943+06-Oct-2008+MW20081006.

Goleman, D. (1998, November–December). What makes a leader? *Harvard Business Review*, 93–102.

Heaney, J. (1997, November 12). Evans blasts police bill as union "power play." *The Boston Herald*, p. 18.

Hub patrolmen don't back Evans. (2002, October 17). *The Boston Herald*, p. 22.

Judge finds for officers beaten by fellow cops. (1998, November 25). *The Associated Press*, state and local wire.

Kobe, L.M., R. Reiter-Palmon & J.D. Rickers (2001). Self-reported leadership experiences in relation to inventoried social and emotional intelligence. *Current Psychology*, 20 (2), 154–163.

Lakshmanan, A.R. (1994, January 15). Evans acting quickly as acting police head. *The Boston Globe*, p. 13.

Lord, R.G., C.L. Devader & G.M. Alliger (1986). A meta-analysis of the relationship between personality traits and leadership perceptions: An application of validity generalization procedures. *Journal of Applied Psychology*, 71, 402–410.

Mallia, J. (1994, April 23). Evans' crime bill is cops key: Neighborhood policing nearing reality. *The Boston Herald*, p. A1.

Marantz, S. (2002, December 25). City nixes new cops in freeze on hiring. *The Boston Herald*, p. 1.

Martinez, J. (2002, March 27). Fed judge rules he'll review race-based cop promotions. *The Boston Herald*, p. 16.

McGrory, B. (1996, July 9). Local officials join Clinton in outlining youth crime bill. *The Boston Globe*, p. 3.

Menendez, J. (1998, August 13). Evans: Keeping city kids busy is key. *The Boston Herald*, p. 27.

More support for embattled police commissioner. (2002, September 20). *The Associated Press*, state and local wire.

Mumford, M.D., S.J. Zaccaro, M.S. Connelly & M.A. Marks (2000a). Leadership skills: Conclusions. and future directions. *Leadership Quarterly*, 11 (1), 155–170.

_____, S.J. Zaccaro, S.J. Harding, D. Francis, T.A. Jacobs & E.A. Fleishman (2000b). Leadership skills for a changing world: Solving complex social problems. *Leadership Quarterly*, 11 (1), 11–35.

_____, M.A. Marks, M.S. Connelly, S.J. Zaccaro & R. Reiter-Palmon (2000c). Development of leadership skills: Experience and timing. *Leadership Quarterly*, 11 (1), 87–114.

Negri, G. (1994, February 14). Evans' leadership skills were honed by demands of his bereaved family. *The Boston Globe*, p. 18.

Noonan, E. (1999, October 28). Officer accused in Cox beating terminated from force. *The Associated Press*, state and local wire.

Paul F. Evans (n.d.). Retrieved July 10, 2003, from http://policetraining.rio.maricopa.edu/staff/paul_evans.shtml.

Pernick, R. (2001). Creating a leadership development program: Nine essential tasks. *Public Personnel Management*, 30 (4), 10–17.

Ranalli, R. (2002, July 1). Girl's shooting spurs bad memories: Many fearing a rise in crime. *The Boston Globe*, p. B6.

Rosenwald, M.S. (2002a, August 22). Violent crime at 31-year low in Boston; commissioner acknowledges residents' fears. *The Boston Globe*, p. A1.

_____ (2002b, September 29). Evans struggles with responsibility to two worlds: New plans show dilemma in protecting police, public. *The Boston Globe*, p. B1.

_____ & A. Estes (2004, February 20). Historic moment for police. *The Boston Globe*, A1.

Sennott, C.M. (2004, May 20). An American police chief in London. *The Boston Globe*, D1.

Sheriff, D.R. (1968). Leadership skills and executive development: Leadership mythology vs. six learnable skills. *Training and Development Journal*, 22 (4), 29–34.

Stand with Evans, a leader. (1998, January 8). *The Boston Herald*, p. 24.

Szaniszlo, M. (2002, September 21). Black leaders back police commissioner; union sticks to demand for resignation. *The Boston Globe*, p. 4.

Technology Association of Maine. (2009). *Penobscot Bay media welcomes, former commissioner of Boston Police Department, Paul Evans to its public safety team.* Retrieved from technology Association of Maine website: http://www.techmaine.com/node/3388.

Valliancourt, M. (1995, June 25). Police hear 100 ideas on public safety. *The Boston Globe,* p. 19.

Woodlief, W. (2002, September 24). Evans finds the right thing tough to do. *The Boston Herald,* p. 27.

Yammarino, F.J. (2000). Leadership skills: Introduction and overview. *Leadership Quarterly,* 11 (1), 5–9.

Zaccaro S.J., J. Gilbert, K.K. Thor & M.D. Mumford (1991). Leadership and social intelligence: Linking social perceptiveness and behavioral flexibility to leader effectiveness. *Leadership Quarterly,* 2, 317–331.

Career of Leadership
Trait Theory

OVERVIEW OF THE THEORY

According to trait theory, an individual's personal traits, such as intelligence, persistence, self-confidence, initiative, and task knowledge, determine his/her success as a leader. This theory is examined through the example of Chief George Edwards, who led the Detroit Police Department for a short period during the 1960s, as an individual whose entire personality contributed to his leadership of the department.

This theory focuses on the characteristics of great leaders throughout history (Northouse, 2004). Here, theorists attempted to identify specific traits that distinguished leaders from non-leaders. The earliest trait theorist may have been Plutarch (c. A.D. 100), who compared the characteristics and behaviors of Greek and Roman leaders (Bass, 1990).[1]

Early trait theory/"Great Man" theory held that leadership characteristics were innate, fixed, and relevant to all situations (Hollander & Offermann, 1993). In the 19th century, the leadership traits included physical characteristics (e.g., height; Bryman, 1992).

Pure trait theory views personality traits as determining leadership in isolation; particular situations are not emphasized (Bass, 1990; Stogdill, 1974). From approximately 1900 to 1940, researchers attempted to establish intrinsic traits differentiating leaders from non-leaders by profiling leaders like Napoleon, Hitler, Gandhi, and Kennedy (Jago, 1982). Sarachek (1968) used characters in Homer's *Iliad* to build archetypal

---

Trait Theory

1. According to trait theory, characteristics of great leaders throughout history determine their success.
2. Early trait theory ("Great Man" theory) posited innate characteristics, fixed and relevant in all situations.
3. Leadership traits in 19th-century trait theory included physical characteristics such as height.
4. Modern applications of trait theory note that personal characteristics are important, as are situations.
5. Psychoanalytical spin on trait theory states that leaders are like artists (Zaleznik, 1977).

---

[1]Plutarch can also be interpreted as a situational theorist, as each leader he describes from Greece had a counterpart in Rome in a similar political context (Bass, 1990).

leadership qualities: Agamemnon (justice, judgment), Nestor (wisdom, counsel), Odysseus (shrewdness, cunning), and Achilles (valor, action; Stogdill, 1974).

MODERN APPLICATIONS OF TRAIT THEORY

Today, trait theory emphasizes that a combination of personal characteristics contributes to successful leadership; however, situations may also be important (Stogdill, 1974). Despite many scholars declaring trait theory dead, it continues to appear in modern leadership materials (Baruch & Lessem, 1997; Jago, 1982; Owens, 1973). "Trait theory is important, because it seeks a framework based on the qualities of exceptional leaders in terms of personal characteristics. This framework can be used to identify potential leaders" (Rowley, 1997, pp. 80–81).

Leaders are like artists. Zaleznik (1977) revitalized trait theory, giving it a psychoanalytic spin. Zaleznik (1977) contrasted managers and leaders based on personality characteristics. Managers were bureaucratic short-term thinkers embedded in routine and inflexibility, while leaders have characteristics that help them to develop new ideas for long-standing problems making them ideal at shaping the bigger picture of the organization. They are "temperamentally disposed to seek out risk and danger" (Zaleznik, 1977, p. 72). They relate to others in an intuitive and empathetic manner and are often emotionally intense. "Leaders are like artists and other gifted people who often struggle with neuroses" (Zaleznik, 1977, p. 75). Zaleznik (1977) emphasized talent over learned skills; individuals with such talent need mentorship in order to actualize their potential. "The only sure way an individual can interrupt reverie-like preoccupation and self-absorption is to form a deep attachment to a great individual or other person who understands and has the ability to communicate with the gifted individual" (Zaleznik, 1977, p. 75).

Leadership traits can be acquired and learned by example. A study by Bennis and Nanus (1985) of 90 leaders led to the emergence of four major skills pertaining to leadership:

1. *Attention through vision.* Leaders are visionaries, have an agenda, and are results-oriented.
2. *Meaning through communication.* Leaders can project meaning to the members of an organization. They articulate what has previously been unsaid and construct images and metaphors to mobilize attention to a new outlook or perspective.
3. *Trust through positioning.* Leaders operate with integrity and buy into their own ideals, fostering trust with their subordinates.
4. *Deployment of self through positive self-regard.* Leaders accept people as they are, approach relationships in terms of the present, treat those closest to them with the same respect as casual acquaintances, trust others, and do not need constant approval and recognition from others.

Bennis and Nanus (1985) also identified several myths about leadership: 1) leadership is a rare skill; 2) leaders are born that way; 3) leaders are charismatic; and 4) leadership exists only at the top of organizations. Likewise, the Bennis and Nanus approach seems to democratize trait theory and encourage people to cultivate their own leadership characteristics.

Certain traits are preconditions to leadership. According to Kirkpatrick and Locke (1991) leadership traits are merely preconditions and do not guarantee success. Kirkpatrick and Locke (1991) identify six core traits that are preconditions to effective leadership: drive, desire to lead, honesty and integrity, self-confidence, cognitive ability, and knowledge of the business. If these traits are possessed, a potential leader can actualize them by implementing the following actions: skills development (as in specific capacities), vision creation, and implementation of the vision.

Managers vs. Leaders

Managers

- Are bureaucratic short-term thinkers embedded in routine and inflexibility (Zaleznik, 1977).
- Join forces to sell goods and services (Bennis & Nanus, 1985).

Leaders

- Develop new ideas for long-standing problems, shape the bigger picture of the organization, seek out risk and danger, and are often emotionally intense (Zaleznik, 1977).
- Work to create a real change (Bennis & Nanus, 1985).

According to Kirkpatrick and Locke (1991), managers should choose employees with the aforementioned traits because these employees possess the characteristics from which leaders can be built. Trait assessment tests can be used to help determine the appropriate staff members. They should then be offered training and experience to facilitate actualization of their leadership traits. "Leaders do not have to be great men or women by being intellectual geniuses or omniscient prophets to succeed, but they do need to have the 'right stuff' and this stuff is not equally present in all people" (Kirkpatrick & Locke, 1991, p. 58).

Leadership traits are in the eyes of the beholder. Calder (1977) posits that leadership is built on traits but that the traits need to exist only in the perceptions of others for the leader to be successful. People define others as either possessing or not possessing leadership traits. From this approach (aka. the attribution theory of leadership), the followers' perceptions of leadership qualities determine whether the leader will be effective (Jago, 1982).

Leadership Traits Can Be Learned

1. ***Attention through vision.*** Leaders have an agenda and are results-oriented.
2. ***Meaning through communication.*** Leaders have the capacity to project/articulate meaning.
3. ***Trust through positioning.*** Leaders operate with integrity and buy into their own ideals.
4. ***Deployment of self through positive self-regard.*** Leaders project acceptance, respect, and trust.

(Bennis & Nanus, 1985)

Certain Traits Are Preconditions

1. Drive
2. Desire to lead
3. Honesty and integrity
4. Self-confidence
5. Cognitive ability
6. Knowledge of the business

If you have these, all you need is skills, vision, and implementation.

(Kirkpatrick & Locke, 1991)

Leadership Traits Are in Eyes of Beholder

The followers' perceptions of leadership qualities determine whether the leader will be effective or not. This is also known as the attribution theory of leadership.

(Calder, 1977)

SOCIAL SCIENTIFIC EVALUATION OF LEADERSHIP TRAITS

Stogdill (1974) evaluated 124 trait studies; the subjects for which were obtained by observing behaviors of potential subjects in group situations, doing peer reviews, nominating subjects using researchers, selecting people occupying positions of leadership, and analyzing case history data. The findings of Stogdill's study are summarized in Northouse (2004). However, a more complete list of traits associated with leadership according to Stogdill's 1948 study is as follows: capacity (intelligence, alertness, verbal facility, originality, judgment), achievement (scholarship, knowledge, athletic accomplishments), responsibility (dependability, initiative, persistence, aggressiveness, self-confidence, desire to excel), participation (activity, sociability, cooperation, adaptability, humor), status (socioeconomic position, popularity), and situation (mental level, status skills, needs and interests of followers, objectives to be achieved).

As evidenced above, although traits are important, Stogdill also factored in situational variables. He states, "[T]he pattern of personal characteristics of the leader must bear some relevant relationship to the characters, activities and goals of the followers" (pp. 63–64).

According to Mann's (1959) study of 1,400 studies on personality and leadership, the best predictor of leadership success is intelligence. Masculinity and dominance were also found to be positively correlated with leadership. These traits were also found to be significant in a study by Lord, DeVader, & Alliger (1986). However, the two studies disagree as to the correlation of extroversion on leadership. Mann (1959) found it to be significant while Lord et al. (1986) found it to be not significant. Mann (1959) reviewed studies that measured the concept in four ways: observation of individuals' leadership traits by researchers, peer review of leaders' traits, use of status in an organization as the criterion for leadership, and leaders' self-ratings of their traits. The median correlations were low; thus the findings are relatively weak even when statistically significant. However, the methodology in the Lord et al. (1986) study showed more statistical significance for the traits of intelligence, masculinity, and dominance using the same data.

Jenkins (1947; as cited in Stogdill, 1974) reviewed 74 military studies and found that although leaders exhibit a variety of abilities and characteristics that are superior to followers,

Social Scientific Evaluation of Leadership Traits

1. Capacity
2. Achievement
3. Responsibility
4. Participation
5. Status
6. Situation

(Stogdill, 1974)

Social Scientific Evaluation of Leadership Traits

1. Drive for responsibility
2. Completion of tasks
3. Vigor and persistence
4. Originality in problem solving
5. Social initiative
6. Self-confidence
7. Sense of personal identity
8. Acceptance of consequences
9. Tolerance of frustration and delay
10. Ability to influence others' behavior
11. Capacity to structure social interaction

(Bass, 1990)

a pattern of specific traits could not be discerned. From this, he deduced that a trait theory of leadership could not be supported. Rather, he postulates that leadership is situation-specific (Stogdill, 1974).

Bass (1990) states that the following traits are strongly correlated with leadership: drive for responsibility, completion of tasks, vigor and persistence, originality in problem solving, social initiative, self-confidence, sense of personal identity, acceptance of consequences, tolerance of frustration and delay, ability to influence others' behavior, and capacity to structure social interaction. However, Bass (1990) also argues that there is an interaction effect between the situation and the traits possessed by leaders. "There is no overall comprehensive theory of the personality of leaders. Nonetheless, evidence abounds that particular patterns of traits are of consequence to leadership, such as determination, persistence, self-confidence and ego strength" (Bass, 1990, p. 87).

Krimmel and Lindenmuth (2001) researched the desirable and undesirable performance indicators of police chiefs by examining the attitudes of 250 municipal managers regarding police chiefs under their control. The managers answered questions about 45 different leadership indicators of their police chiefs; the questions primarily focused on a list of traits such as "responsible," "shares information," "focused," and "sophisticated" (Krimmel & Lindenmuth, 2001). The list of traits is broken down into six subsets (Krimmel & Lindenmuth, 2001):

1. *Sad.* This refers to responses that demonstrate that the police chief is defensive, dissatisfied, lonely, vulnerable, and unable to delegate authority.
2. *Upset.* This refers to responses indicating that the chief is insensitive, arrogant, and betrays trust; is inappropriately ambitious; rules with an iron fist; has boot camp values; and denies responsibility.
3. *Calm.* This subset refers to chiefs who are fair-minded, sincere, reasonable, comfortable with uncertainty, competitive, conscientious, and able to focus on the whole problem, and who recognize the value of disequilibrium.
4. *Sharing.* This subset reflects chiefs who are people persons, collaborators, information sharers, providers of clear visions, developers of new patterns of relationships, motivators of employees, and outgoing people.

5. *Scout.* This refers to the subset of variables reflecting that the chief takes responsibility, is responsible, selects good subordinates, is appropriately ambitious, is well organized, selects and maintains goals, and properly maintains activities.
6. *Bond* (short for James Bond). This subset reflects chiefs who are self-controlled, cautious, reserved, capable, secure, and intelligent.

Krimmel and Lindenmuth (2001) found that police chiefs promoted from within the department tended to fit into the calm, sharing, or scout subsets. Chiefs hired from the outside were more likely to be categorized in the upset subset. Graduates of the FBI National Academy were more likely to be rated in the calm, sharing, scout, or bond subset. Chiefs possessing (at best) high school diplomas were more likely to be rated as having leadership problems while police chiefs with better education were more likely to be evaluated as belonging to the sharing or scout subset. The type of community policed had no effect on leadership indicators. However, police departments that were unionized were significantly more likely to evaluate police chiefs as belonging to the calm, sharing, or scout subset. In summary, police chiefs who have some college credits, have graduated from the FBI National Academy, and have been promoted from within the department are the least likely to be assessed as having leadership problems (Krimmel & Lindenmuth, 2001).

TRAIT THEORY AND THE CASE OF CHIEF GEORGE EDWARDS

Trait theory posits that an individual's characteristics determine his/her success as a leader. This theory is based on the assumption that leaders are born not made. Traits identified as leadership qualities include motivation, drive, and persistence. These three qualities are the best indicators of the leadership skills chosen in this example.

However, as important as personal characteristics are to trait theory, they are also dependent on the situations that leaders encounter. In other words, one might have the leadership traits or characteristics but never encounter the right situations to be able to make the best decision or judgment and apply those traits. Thus, an interesting spin on trait theory offers the possibility that traits are acquired, not ascribed, as a result of the variety of situations that individuals encounter on their paths to becoming leaders.

CAREER HIGHLIGHTS

Trait Theory

Chief George Edwards
Detroit Police Department, Michigan

George C. Edwards, Jr., was born in Dallas, Texas, in 1914. Edwards received his bachelor's degree from Southern Methodist University in 1933 and his master's degree in literature from Harvard in 1934 ("George C. Edwards, Jr., collection," n.d.).

From 1934 to 1935, he worked as the secretary of the League of Industrial Democracy, traveling around the country as a labor activist. He relocated to Detroit in 1936, working in an automobile plant while writing a book (never published). He participated in a sit-down strike in 1937 at the plant and was sentenced to 30 days in jail for defying a court order to disperse ("George C. Edwards, Jr., banned," 1995).

In 1937, he worked as the secretary-director of the Detroit Housing Commission. He was elected to the Detroit Common Council in 1941, subsequently serving four terms. From 1943 to 1946, Edwards was a lieutenant in the U.S. Army during which time he investigated Japanese war crimes in the Philippines ("George C. Edwards, Jr., 80," 1995; Stolberg, 1998). He then became the national director of the United Auto Workers welfare department ("George C. Edwards, Jr., 80," 1995).

Edwards earned his law degree from the Detroit College of Law in 1944 and subsequently entered private practice. In 1949, he unsuccessfully ran for mayor of Detroit ("George C. Edwards, Jr., collection," n.d.). Edwards was appointed as a judge for Wayne County's Probate Court and revised procedures for handling juveniles. He implemented preliminary hearings within 24 hours of the apprehension of a juvenile and sped up the process in subsequent stages. These reforms came in response to a system in which juveniles had been held in limbo for months (Stolberg, 1998). In 1954, Edwards was elected to the Wayne County Circuit Court in Michigan, followed by his 1956 appointment to the Michigan Supreme Court. He left this post in 1962 to serve as police commissioner in Detroit ("George C. Edwards, Jr., banned," 1995; "Judge G. C. Edwards," 1995).

After serving as commissioner (discussed below), Edwards served on the 6th Circuit U.S. Court of Appeals from 1963 to 1985, when he semi-retired and took a reduced caseload. He was chief judge of the circuit from 1979 to 1983. He is remembered for writing the opinion in a 1971 case that established that wiretapping without a warrant was unconstitutional. The opinion was upheld by the U.S. Supreme Court and was later applied to the Watergate proceedings ("Deaths elsewhere," 1995). The case stemmed from the conviction of Lawrence Robert Plamondon, a White Panther party member, on charges that he bombed a CIA office in Ann Arbor, Michigan, in 1969 ("George C. Edwards, Jr., banned," 1995).

Edwards served as commissioner of the Detroit Police Department from 1962 to 1963. Mayor Jerome P. Cavanaugh chose Edwards as police commissioner because of Edwards' record of working for racial integration and harmony (Stolberg, 1998). On the bench, he had been particularly sensitive to constitutional issues associated with racially discriminatory enforcement of the law. Edwards' record was especially suitable given the controversy surrounding a crackdown in Detroit by the previous police chief, William D. Hart. In 1960, in response to the murders of two white women at the hands of a black man, Hart declared a state of emergency and ordered a roundup of street criminals. The black community largely believed the crackdown to be an excuse for police officers to intimidate and harass blacks. Community leaders were becoming increasingly outspoken about the racial divide in the city and viewed the police force as a hostile white enemy (Stolberg, 1998). Furthermore, in 1943, after years of increasing racial tensions, the city erupted into a race riot.

Edwards reportedly deliberated about whether to take the job as commissioner. As a Michigan Supreme Court judge, he already had a high-profile job. However, being commissioner would allow him to explore the practicalities of what already intrigued him: how to balance constitutional safeguards with law enforcement (Stolberg, 1998). In a letter to his son about his decision, he wrote, "I guess I couldn't resist trying to figure out whether it is possible to make the Constitution a living document in one of our great cities" (Stolberg, 1998, p. 36). In response to Edwards' appointment, numerous black civic leaders praised Mayor Cavanaugh. Reaction among Detroit's white residents was reportedly divided. Upon taking the job in 1962, Edwards laid out his main goals for his tenure as commissioner: improve racial equality in the enforcement of the law, target the Mafia, increase minority recruiting in the police department, and develop community-oriented policing (Stolberg, 1998).

Improve racial equality in law enforcement. In 1962, the city of Detroit was one-third black. Edwards' first reform move for racial equality was to overhaul the department's traffic division, which had a bad reputation in the black community for harassment. In an effort to reach out to the black community, he put pressure on the deputy commissioner who headed the division to treat black motorists with respect (Stolberg, 1998). Despite skepticism about whether his overtures to the deputy commissioner would make a difference, within a few weeks, Edwards began receiving letters from black citizens and civic leaders thanking him for the change in police attitudes (Stolberg, 1998).

His first race-related crisis occurred on January 28, 1962, when Willie Daniels, a black man accused of threatening a woman with a gun, was beaten by police officers when he was apprehended in his home (Stolberg, 1998). Daniels sued the city for his injuries (bruising and swelling on his wrists and stomach), which were verified by a doctor. Meanwhile, Edwards initiated an internal investigation, including a trial board consisting of Edwards and top brass. Despite ongoing complaints of brutality, an investigation had not been initiated since 1957. After conducting the internal trial, the board found the accused officers not guilty, though Edwards made it clear the decision was not unanimous because he believed otherwise. Edwards stated that from the trial he realized that "police morale was a code name for a police officer's obligation to support another police officer, regardless of regulation or law unless he was so far out of bounds that saving him was impossible" (Stolberg, 1998, p. 159 [quoting from Edwards' unpublished manuscript about his commissionership]). Edwards subsequently changed departmental policy, requiring a supervisor to go to every location at which a gun was reported in hopes of decreasing brutality and increasing accountability.

Another brutality issue cropped up near the end of Edwards' tenure. Cynthia Scott, a known prostitute and robber, threatened an officer with a knife while being escorted to a squad car (Stolberg, 1998). After she cut the officer, she ran off. The officer's partner gave chase, and she turned on him with the knife as well. He told her to stop, but she stepped away; the officer subsequently shot her in the back, killing her (Stolberg, 1998). The case was a public relations disaster because the black community felt it was an example of brutality. The incident sparked picketers at the City-County Building. Edwards conducted his own internal inquiry, ultimately agreeing with the attorney general and district attorney, who declined to take action against the officer who shot Scott. Edwards believed the officer was justified because there was little doubt that he was afraid for his life at the time of the shooting (Stolberg, 1998). The case did damage to the black community's perception of the police in spite of the efforts of reform that Edwards had previously been praised for (Stolberg, 1998).

Despite Edwards' commitment to reform and his sensitivity to the black community, he stood against civilian review boards (Stolberg, 1998). He believed that if he made it clear that excessive use of force would not be tolerated, civilian oversight would be unnecessary. On this point, many black leaders disagreed, saying that true reform required civilian review boards (Stolberg, 1998). However, in 1962, Edwards gave the Community Relations Bureau within the department the responsibility for investigating brutality complaints, setting up a specific team that he felt would be unbiased (Stolberg, 1998).

Edwards' handling of security at potentially racially charged events in Detroit also received praise for decreasing tensions and ensuring public safety (Stolberg, 1998). On June 10, 1962, Nation of Islam leader Malcolm X spoke before 3,500 people in the city. Edwards attended the event, although he did not explicitly support the group. Instead, he told the press that the group had a right to meet in the city and be treated with respect as long as they did not pose any threat to lives or property. Edwards had 900 officers on standby in the event of a disturbance, but they were ultimately unneeded. He took a similar approach to the Freedom March staged by Martin Luther King, Jr., on June 23, 1963. Final crowd estimates were between 125,000 and 250,000 people, including Edwards who marched with them. The only trouble reported to the police that day was a broken window and a disoriented man in the middle of a street; there were no violent incidents (Stolberg, 1998).

Target the Mafia. While commissioner, Edwards fought against the influence of the Mafia in the city, particularly seeking to disrupt gambling operations (Stolberg, 1998). He was a proponent of the alien conspiracy theory, which holds that mob bosses are connected under a national umbrella (Stolberg, 1998). This was a common conspiracy theory at the time, as 58 mob bosses across the nation were found meeting in Apalachin, New York, in 1957. That same year, the McClellan Commission was set up by the Senate to explore the activities of organized criminals infiltrating labor unions (Stolberg, 1998).

Edwards began his antimob crusade by forcing the shutdown of the Lesod gambling club on the Lower East Side of the city (Stolberg, 1998). He built a case against the establishment via police surveillance and interviews of customers; the mob eventually conceded. He also implemented a policy to discourage other clubs from operating in Detroit by passing an ordinance (with the city council) prohibiting clubs to be open between 2 a.m. and 8 a.m. unless they agreed to a police investigation of their activities (Stolberg, 1998). Edwards' biggest antimob action was the raid on the Gotham Hotel, orchestrated in conjunction with the IRS: 41 people were arrested, $60,000 in cash was seized, and 120,000 betting slips were found. In addition, Edwards led a successful investigation against mob boss Giacamo Giacalone using wiretaps (Stolberg, 1998).

Increase minority recruitment. Edwards believed that diffusing racial tensions required minority recruitment of police officers. At the time of his appointment, there were only 136 black police officers in a force of 4,300 (which equaled 3 percent black; Stolberg, 1998). This occurred despite the fact that blacks made up 28% of the applicant pool, similar to their representation in the general population. Furthermore, no black officer had achieved the rank of inspector. A report by Commissioner Director Richard Marks found that discrimination in hiring practices was routine (Stolberg, 1998). Higher standards on physical, mental, and medical tests were applied to black applicants. During his tenure, Edwards promoted blacks, including a black officer as inspector, and began a media campaign to recruit minorities. However, he was able to make only modest gains. By 1967 (four years after Edwards left the department), there were only 191 officers in contrast to 4,049 white officers (Stolberg, 1998).

Develop community-oriented policing. Edwards is credited with envisioning community-oriented policing as an answer to police-community relations, including race relations (Stolberg, 1998). He hired more officers and strategically placed them in high-crime areas at peak times. He also fostered relationships with neighborhood block clubs designed to ward off crime (Stolberg, 1998).

Edwards also continued police-community meetings that had begun under the previous commissioner (Stolberg, 1998). In particular, on June 22, 1962, Edwards organized a conference among civilians and civic and religious leaders on juvenile delinquency in order to encourage citizen cooperation in combating the problem. The highlight of the meeting was an appearance by Raymond Burr (Stolberg, 1998). On November 1, a follow-up conference with public and private school officials was held to discuss local juvenile delinquency concerns (Stolberg, 1998). Edwards also received federal funding for a program called Community Action for Detroit Youth, which researched and developed initiatives to combat juvenile delinquency at the community level (Stolberg, 1998).

Despite the above reforms that garnered community support, Edwards did not enjoy popularity among the rank and file (Stolberg, 1998). At the announcement of his appointment as commissioner, officers largely disapproved because of his judicial record of liberalism (Stolberg, 1998). In his first few weeks on the job, Edwards made an effort to visit various station houses and talk to as many officers as possible in order to reach out and gain support (Stolberg, 1998). Yet, despite this attempt, officers often subtly sabotaged his policies. For example, as part of a community-policing initiative, Edwards asked precinct commanders to meet with school officials after responding to gang fights. When this did not happen two weeks later, he ordered them to do so and to provide him with written reports (Stolberg, 1998). In another incident, department leaders misreported crime statistics in order to discredit Edwards' approach to crime fighting (Stolberg, 1998).

Despite Edwards' negotiation of a pay increase for officers, on April 16, 1962, department leaders called a meeting to discuss their dissatisfaction with him (Stolberg, 1998). Most of the complaints involved low morale and a sense that their ability to do their job was hindered by various reform policies (Stolberg, 1998). Edwards never commanded the respect of most of the rank and file; however, his primary aim was not to win their support but to win that of the public (Stolberg, 1998). As Edwards stated:

> The issue of whether there will be civilian control of the police . . . is one of the most critical issues of our time. . . . The person who functions as the executive head of the police department should have his eye on the civilian need for his department, not whether he is popular with his rank and file. (Blonston, 1968, cited in Stolberg, 1998, p. 292)

Chief Edwards' experience shows motivation, drive, and persistence. Edwards was also an individual who was exposed to a myriad of situations and experiences throughout his career, especially during the times when he toured the country as a labor activist. There is an old adage: Nothing gives you a better education than travel. There is much to be said about this wisdom with regard to exposure to situations and learning of leadership traits. Changing work environments and even serving time in prison for doing the right thing certainly contributed volumes to Edwards' learning experience.

When one looks at Edwards' life, it is hard to ignore his drive to be successful whatever the endeavor. No matter how unrelated his fields of interest, they share a common thread: Edwards' motivation to complete his task well. It was precisely, this motivation, drive, and persistence,

paired with the experiences offered by his eclectic career, that made Edwards a suitable candidate to be chosen as the chief of the Detroit Police Department. Edwards did not need the position to advance his political career, as he already had a high public profile. He did not need another challenge, since he already had quite a full plate of challenges. However, it appears that, above all, his motivation to take on the position was a factor of a pronounced leadership trait—the desire to do the right thing.

It is natural to try to consider deeper motivations when trying to understand situations in which we (or others) have become involved that at least superficially seem to bear little, if any, benefit. Frequently we are even asked why we do the things we do or why we involve ourselves in a predicament that might destroy our past accomplishments and bring little if any glory to our future. Sometimes actions that appear to make no sense whatsoever are motivated by leadership qualities and traits more than by some irrational drive. Occasionally it is a desire and drive for honesty and integrity, vigor and persistence, and a sense of personal identity as well as an overall burning desire to restore some balance of injustice and to do the right thing.

Chief Edwards' tenure as police commissioner, paired with his life experience in other areas, illustrates a true example of the trait theory of leadership. The way he handled sensitive situations, race-related problems, and the introduction of innovative community-oriented policing (two decades prior to the formal introduction of such concepts by social scientists)—these are just a few examples of a leader who needed to have the right characteristics that enabled him to do the right thing in the politically and socially volatile world of policing in the 1960s.

After leaving the Detroit Police Department, George Edwards continued to be a member of the Advisory Council of Judges-National Council on Crime and Delinquency until 1975 (History, n.d.). In 1963, he received the Citizen of the Year Award from the *Michigan Chronicle,* and along with his wife, received the Amity Award for civic achievement (History, n.d.). Between 1965 and 1971, he was the Chairman of the Committee on Administration of Criminal Law, and a member of the Advisory Committee on Criminal Rules; both part of the United States Judicial Conference (History, n.d.). Edwards was also a member of the National Commission on Reform of Federal Criminal Laws between 1967 and 1971 (History, n.d.).

Between 1966 and 1971, Edwards received two awards; the American Society of Criminology August Vollmer Award (1966) and the Association of Federal Investigators Judiciary Award (1971) (History, n.d.). During this time, he also received a Doctor of Laws from William Mitchell College of Law (1969; "History," n.d.).

In the 1970s, he moved to Cincinnati where he participated in the decision in the Detroit busing case ("History," n.d.). Edwards eventually became Senior Circuit Judge of the Sixth U.S. Circuit Court in January 1985 ("History," n.d.). Edwards died ten years later in 1995 ("History," n.d.).

Exercise

The police chief detailed in this chapter has been examined through the prism of trait theory. Using what you have learned from other theories in this book or from the summaries (as indicated in the Introduction) you have read thus far, it is possible to consider his approach through the lens of different theories. What other theories can you find applicable to this chief? Are there lessons to be learned from those applications, or do those theories indicate alternative paths or approaches that might have yielded a more successful outcome?

References

Baruch, Y. & R. Lessem (1997). The spectral management inventory—A validation study. *Journal of Managerial Psychology,* 12, 365–382.

Bass, B.M. (1990). *Bass and Stogdill's handbook of leadership: A survey of theory and research.* New York: Free Press.

Bennis, G., & Nanus, B. (1985). *Leaders: The strategies for taking charge.* New York: Harper & Row.

Bryman, A. (1992). *Charisma and leadership in organizations.* London: Sage.

Calder, B.J. (1977). An attribution theory of leadership. In B.M. Staw & G.R. Salancik (Eds.), *New directions in organization behavior* (pp. 179–204). Chicago: St. Claire Press/London: MacMillan. (Paperback edition by Meridian Books, New York, 1962.)

Deaths elsewhere. (1995, April 10). *Chicago Daily Law Bulletin,* p. 3.

George C. Edwards, Jr., banned secret wiretaps. (1995, April 9). *Chicago Sun-Times,* p. 61.

George C. Edwards, Jr., collection. (n.d.). Retrieved August 26, 2003, from www.reuther.wayne.edu/collections/hefa_10.htm.

George C. Edwards, Jr., 80, was chief appeals court judge. (1995, April 9). *The Record,* p. N7.

History of the Sixth Circuit: George Clifton Edwards, Jr. (n.d.). Retrieved from http://www.ca6.uscourts.gov/lib_hist/courts/circuit/judges/edwards/gce-bio.html.

Hollander, E.P. & L.R. Offermann (1993). Power and leadership in organizations. In W.E. Rosenbach & R.L. Taylor (Eds.), *Contemporary issues in leadership* (pp. 62–86). Boulder, CO: Westview Press.

Jago, A.G. (1982). Leadership: Perspectives in theory and research. *Management Science,* 28 (3), 315–336.

Judge G.C. Edwards dies at 80; rules against secret wiretaps. (1995, April 10). *The New York Times,* p. B14.

Kirkpatrick, S.A. & E.A. Locke (1991). Leadership: Do traits matter? *The Executive,* 5, 48–60.

Krimmel, J.T. & P. Lindenmuth (2001). Police chief performance and leadership styles. *Police Quarterly,* 4 (4), 469–483.

Lord, R.G., C.L. DeVader & G.M. Alliger (1986). A meta-analysis of the relation between personality traits and leadership perceptions: An application of validity generalization procedures. *Journal of Applied Psychology,* 71, 402–410.

Mann, R.D. (1959). A review of the relationship between personality and performance in small groups. *Psychological Bulletin,* 56, 241–270.

Northouse, P.G. (2004). *Leadership: Theory and practice* (3rd ed.). Thousand Oaks, CA: Sage.

Owens, J. (1973). What kind of leader do they follow? *Management Review,* 62, 54–57.

Rowley, J. (1997). Academic leaders: Made or born? *Industrial and Commercial Training, 29,* 78–84.

Sarachek, B. (1968). Greek concepts of leadership. *Academic Management Journal,* 11, 39–48.

Stogdill, R.M. (1948). Personal factors associated with leadership: A survey of literature. *Journal of Psychology, 25,* 35–71.

_____(1974). *Handbook of leadership: A survey of theory and research.* New York: Free Press.

Stolberg, M.M. (1998). *Bridging the river of hatred: The pioneering efforts of Detroit police commissioner George Edwards.* Detroit: Wayne State University Press.

Zaleznik, A. (1977, May–June). Managers and leaders: Are they different? *Harvard Business Review,* 55 (3), 67–68. (Reprinted in 1992, *Harvard Business Review,* 70 (2), 126–135.)

13

Into the Future
The Big Hairy Audacious
Goal and Catalytic Mechanisms

Nearly 200 years ago, leading police reformers advocated removing the police from politics and politics from the police, in other words, to treat this occupation as a true profession. This task, however difficult to achieve, remains weighty and important. The question resembles the proverbial chicken and the egg: Do we as a society owe it to democracy to maintain a strong and professional police force, or is society democratic because of professional policing?

A good overview of leadership theories must end with Jim Collins's leadership theory, the most relevant approach to socially complex and politically impaired police organizations. Although Collins, similar to other researchers who contributed to the development of leadership theories, did not look specifically at police environments to identify the patterns of behaviors and structures, his approach to leadership is particularly relevant for police organizations. It offers insight into how to turn entities fraught with problems and unachievable goals into progressive and fully operational organizations. Despite the fact that his approach has more of a management than purely leadership slant, it speaks to leadership practice and is easily translated into the realities of police life.

Jim Collins's leadership theory is part of his general management theory outlined in his two books, *Built to Last* (1994 [with J. Porras]) and *Good to Great* (2001). In general, Collins advocates that companies set ambitious goals. Leaders should act out their organization's core values and goals rather than honing in on their own self-interest. Maximizing profit is not as effective a goal as maintaining core values. Patience is also a major asset, as turnarounds in an organization can take up to seven years (Lenzner, 2003). Collins's leadership theories are often described as "reality-based" (Bryant, 2003, p. 50). These theories are described and illustrated with examples of some of the police chiefs depicted in this book as well as generally applicable reality-based situations from everyday policing. Finally, the "Big Hairy Audacious" goal of the future police leadership is illustrated with the introduction of the female police leadership, so frequently categorized as "different" or "less competent" than the male's approaches. By introducing three powerful and charismatic female police chiefs, while illustrating their contributions to the police profession, this author hopes to translate the "Big Hairy Audacious" goal of accepting female leadership in police profession not as an anomaly that needs to break the "glass ceiling" but as a very much desired integration of a womanpower into what has been considered for thousands of years a male domain and sadly, still in the years 2011, appears to generate more skepticism than applaud.

ANTICHARISMATIC LEADERSHIP

In *Good to Great* (2001), Collins and his team of 21 researchers sampled 11 large companies that were particularly successful on certain economic indicators, including high cumulative stock returns over a 15-year period and average returns of over three times the market rate. As a comparison, Collins then selected six companies that performed poorly on the same indicators. One of the major findings was that charismatic leadership was negatively correlated with successful businesses.

Collins believes that charismatic and transformational leadership is a dead end for businesses. Businesses must overcome the "culture and affliction" of relying on particularly dominant and self-absorbed leaders:

> In the US, we revere the great, egocentric leaders. We have had a generation of leaders that were ambitious first and foremost for themselves and we are now paying the price (Collins as quoted by Pickard, 2003, p. 8).

The main problem with transformational leadership is that the leader's dominant personality overtakes the reality of the organizational context. Collins explains:

> The moment a leader allows himself to become the primary reality people worry about, rather than reality being the primary reality, you have a recipe for mediocrity, or worse. This is one of the key reasons why less charismatic leaders often produce better long-term results than their more charismatic counterparts. (Collins, 2001, p. 72)

Furthermore, celebrities who take over an organization from the outside are also negatively correlated with business success (Bryant, 2003; Collins, 2001).

When one looks at the police leaders of national prominence who were brought in to transform police organizations fraught with problems of corruption and/or general mismanagement, it appears that there is a striking resemblance between Collins's research findings and their level of success. It is true that in times of extreme distress, a charismatic leader will produce extreme results; however, the question remains how long-lasting those extreme results are going to be. It is true that one can reform an organization in the most extreme manner by introducing innovations. Examples include the use of COMPSTAT, as in New York City, or numerous changes in structure and operational deployment, as in New Orleans. Looking at long-term results, the police forces of New York City or New Orleans or Charlotte-Mecklenburg are not very different than what they used to be prior to experiencing and experimenting with the appointment of their charismatic leaders.

When this point was made in classroom discussions at John Jay College of Criminal Justice, where many students are NYPD officers, it inevitably met with objections. Some NYPD students abruptly disagreed with the notion that the NYPD in the year 2004 was the same as in the year 1994, or 1984 for that matter. However, the sad truth is, it is. It is correct to say that the manner of deployment has changed dramatically; the clearance rate is much higher; the overall level of education and training is better; line officers up to ranking police supervisors are more accountable and effective than their counterparts from years ago; equipment is better; and response time is faster. Nevertheless, this superficially attractive force still suffers from an inability to recruit the best and the brightest; also, the attrition rate is high; the pay is low; the morale is low; tensions between supervisors and troops run high; and the sense of pride in wearing this particular uniform is nowhere near what it should be. Although these findings were not empirically researched, they are based on the testimonial evidence and close weekly interactions with hundreds and hundreds of NYPD officers from 2001 to 2004.

A fully functional and transformed organization cannot exhibit dysfunctional qualities. One cannot claim both success and change for the better if employees are dissatisfied or potential quality candidates do not fight for a job offer within the organization. When employees wake up in the morning having to go to work rather than wanting to go to work, one can posit that the organization is as dysfunctional as it was years ago, even factoring in major innovations and directional changes. Charismatic leaders who come and go can change only so much. Police organizations, especially the larger ones, struggle with lack of leadership from the very foundation on up. Producing leaders at the very bottom, working on leadership skills and qualities from the first day at the police academy, will produce true insiders who will eventually deliver what is needed at the top of the police organization.

As Collins acknowledges, leadership effectiveness is more likely when leaders rise from within the ranks:

> Ten out of the eleven good-to-great CEOs came from *inside* the company, three of them by family inheritance. The comparison companies turned to outsiders with *six* times greater frequency—yet they failed to produce sustained great results (Collins, 2001, p. 32).

Insiders are better able to measure progress and stay committed to company goals due to their understanding of and experience within the organizational culture (discussed below; Bryant, 2003). There is one disclaimer or clarification to be made about insider leaders: Being an insider does not immediately qualify one to be an effective leader of any organization, especially not a police one. That is why we need to develop many potential leaders from whom we will be able to pick *the one,* as defined by Collins in his Level 5 leadership approach.

LEVEL 5 LEADERSHIP

Instead of charismatic leadership, Collins advocates what he terms "Level 5 leadership," which is a model of executive capabilities (Collins, 2001). The Level 5 leader is the best possible leader and exhibits the following characteristics:

- Is ambitious but puts the needs of the organization first
- Is modest, self-effacing, and understated
- Is driven by the need to produce sustainable results
- Sets up successors for effectiveness

Additional adjectives used to characterize Level 5 leaders are quiet, reserved, shy, gracious, mild-mannered, and understanding (Collins, 2001).

Level 5 leaders also employ the concept of "the window and the mirror" (Collins, 2001, p. 33). They attribute their successes to luck or factors outside their control while taking responsibility for their failures. In other words, they take stock of their problems by looking within the organization while maintaining humility in times of success.

One criticism of "the window and the mirror" is a question posed by Macht (2001): "[H]ow does one get an accurate assessment of a company's success if its leaders are too modest to talk about it frankly?" (p. 122). For Macht, this approach of focusing on leaders' attributions of luck and outside forces actually represents a missed opportunity to delve deeper into what leaders actually did to achieve their goals. According to Collins, however, "the window and the mirror" approach helps leaders to focus realistically on their organizational problems while not suffering from a lack of humility when times are good:

> There is nothing wrong with pursuing a vision for greatness. After all, the good-to-great companies also set out to create greatness. But unlike the comparison companies, the good-to-great companies continually refined the *path* to greatness (Collins, 2001, p. 71).

Realism is important to Collins's leadership vision. This means fostering a culture of truth in which nothing gets in the way of dialogue with subordinates and a realistic approach to organizational problems. The capacity to actualize a culture of truth rests on having humility and being able to "understand enough to have the answers and then to ask the questions that will lead to the best possible insights" (Collins, 2001, p. 75). It also rests on a paradoxical notion that while confronting the brutal reality, a Level 5 leader also retains faith that the organization will prevail in the end despite difficult circumstances (Collins, 2001).

The game of truth is a dangerous one in today's police organizations, primarily because of the lack of leadership training that needs to be and must be offered at the basic academy level. Too often, police organizations have an overreliance on rank structure and attach ultimate authority to the position one holds. They underdevelop referent power, which stresses and emphasizes authority as a derivative of respect for a supervisor. To develop an environment of truth, one needs to be skilled in many leadership approaches and methods. This skill does not arise from intuition; it comes from years of learning and experimenting. However, this experimentation needs to be part of a structured process of learning and not a function of uncontrolled errors of judgment.

In addition, Collins found no link between executive compensation and company performance, also suggesting the role that humility plays in leadership. Furthermore, flashy technology does not correlate with good business (Macht, 2001). In general, money and hype are considered smoke screens by Collins, who favors leaders with a sense of values and vision. In police environments, in which money and hype are certainly not what the job is all about, this approach to leadership appears to be more relevant than in any other company where Collins might have conducted his research. Although many would say that what attracts them to policing is the prospect of steady employment and salary, it does not take much time on the job to realize that being a police officer is not about money and hype; it is about the war between good and evil and the difference between right and wrong. No matter how romanticized this notion might appear to an outsider reader, this is what police work is all about.

It is unclear whether Level 5 leaders can be trained and developed or whether the traits are innate. Collins has indicated that people can evolve into Level 5 leadership by focusing on the needs of their organization rather than on their self-interests. Pickard (2003) is right when he claims that this core focus is not necessarily fostered by sending people to leadership courses. People need to be introduced to leadership concepts from day one of their learning experience. Just as one will not understand the work of Shakespeare with a kindergartner's comprehension of the English language, it will only be possible to create and develop leadership attitudes through indoctrination of leadership concepts from the onset of one's career.

The Level 5 hierarchical model includes four sublevels, which represent lesser leadership capabilities:

> Level 4—*Effective leader.* "Catalyzes commitment to and vigorous pursuit of a clear and compelling vision, stimulating higher performance standards" (Collins, 2001, p. 20).

> Level 3—*Competent manager.* Organizes subordinates and other resources around the effective and efficient pursuit of goals.

Level 2—*Contributing team member.* Provides individual skills and talents for the achievement of goals and works cooperatively with others in groups.

Level 1—*Highly capable individual.* "Makes productive contributions through talent, knowledge, skills, and good work habits."

It is highly possible that finding an effective leader will be fraught with difficulties, especially in the smaller police departments that do not have the luxury of sifting through hundreds or thousands of employee applications to find *the one.* However, it is strongly recommended to strive for Levels 1 through 3 capabilities to achieve the ultimate formula for an effectively run and fully functional organization.

CHOICE OF THE RIGHT PEOPLE

Collins also believes that leaders succeed by attracting and keeping the right employees. Rather than trying to overhaul and train employees for their positions, organizations should pick suitable, adaptable people from the start. Collins believes these are self-disciplined people willing to go to great lengths to fulfill their responsibilities in the workplace. In fact, the question of who should be in the organization should precede the question of what the organization should be doing. The following three practical rules are suggested:

1. When in doubt, do not hire—keep looking.
2. Act on the need to make a change in personnel.
3. Put your best people to work on the biggest opportunities, not on the biggest problems.

The first and third rules are of critical importance for police environments. When force numbers need to be increased, whether the source of that need is political, social, economic, or something else, police organizations tend to lower their standards and hire the needed numbers. It often appears that there is a total and complete sacrifice of quality for quantity. The greatest misconception of police work states that more police officers equal better police work, which is a true disservice to this noble profession. Better policing will always and under all circumstances be delivered with fewer but better cops rather than with more but lower-quality cops. Hiring for the sake of hiring is a formula for disaster and organizational dysfunction; this leadership rule needs to be hammered into the heads of current (and future) police leaders. The third factor involves a paradox: Managing problems can make an employee good, but to achieve greatness, the best employees should be working on the cutting edge of what the organization is doing (Collins, 2001).

According to Collins, companies tend to view personnel structure as "a genius with a thousand helpers" (Collins, 2001, p. 61). This is commensurate with the charismatic leadership that Collins despises—the notion that there is a genius leader who sets the entire vision, with underlings to carry it out. Instead, by picking the right people, the leader can institute a more democratic workplace that encourages freedom and innovation. This culture, however, should be implemented within a framework. Subordinates should be encouraged to take "disciplined action" (Collins, 2001, p. 124). These actions should reflect the values and culture of the organization:

> The good-to-great companies built a consistent system with clear constraints, but also gave people freedom and responsibility within the framework of that system. They hired self-disciplined people who didn't need to be managed, and then managed the system, not the people (Collins, 2001, p. 125).

Furthermore, Collins suggests that the core values an organization adopts are not as important as the process of soul-searching that occurs in outlining core values. Core values are the glue that holds an organization together and that helps to provide direction (Collins, 2001).

HEDGEHOG RULE OF MANAGEMENT

Leadership requires an understanding of what the organization can *actually* be the best at doing or producing, known as the hedgehog concept. The hedgehog metaphor comes from a Greek proverb: The fox knows many things, but the hedgehog knows one big thing (Lenzner, 2003). Collins has altered the formula, and his "hedgehog" knows three things. According to Collins (2001), leaders must help organizations reflect on:

1. What the organization can be passionate about.
2. What the organization can be the best in the world at.
3. What drives the organization's economic engine.

The first factor involves fulfilling a sense of mission or passion. Subordinates can get excited about the business at hand when it is something employees are personally passionate about:

> The good-to-great companies did not say, "Okay, folks, let's get passionate about what we do." Sensibly, they went the other way entirely: We should only do those things we can get passionate about. Kimberly-Clark executives made the shift to paper-based consumer products because they could get more passionate about them. As one executive put it, the traditional paper products are okay "but they don't have the charisma of a diaper." (Collins, 2001, p. 109)

The potential problem for police organizations in fulfilling passion is that the mission of the police is largely dictated by its function within a democratic society. In other words, enforcing laws, making arrests, and protecting public safety must be on the agenda regardless of whether the agency is passionate about such work. However, the employees can get passionate about the style or type of law enforcement employed by the agency and about special programs that the police can offer the community. Does the agency feel passionate about community policing, team policing, or undercover work? Running a child fingerprinting program? Providing faster 911 response?

The second factor in police work is also dictated by the constraints and definition of policing. Obviously, each police agency needs to be the best at protecting public safety, if they are the best at anything. However, this success can be measured in many ways. Agencies can decide to be the best in crime-clearance rates or victim support functions, for example.

The third factor involves organizations isolating their fundamental moneymaker. According to Collins, less than 10% of organizations understand their true economic indicators (Lenzner, 2003). For public organizations such as police agencies, it may or may not be truly a matter of finances. It could mean having a political sense and the ability to influence local budgetary decision making. It could involve an emphasis on public relations and a sensitivity to the public perceptions of local policing. It could also suggest that leaders should be attempting to obtain grants from outside sources or from the federal or state government for special programs.

BIG HAIRY AUDACIOUS GOALS (BHAGS) AND THE FORMAL AND INFORMAL GOALS OF POLICING

Collins's most famous theoretical formulation is the notion of the "big hairy audacious goal" (BHAG), developed in *Built to Last* (1994 [with J. Porras]). The first step is to start with role models

---

Big Hairy Audacious Goals and Five Catalytic Mechanisms

First Mechanism

- The mechanism can produce desired results in unpredictable ways.

Second Mechanism

- The mechanism distributes power for the benefit of the organization and does not concentrate power at the top.

Third Mechanism

- The mechanism has a "sharp set of teeth" (Collins, 1999, p. 75).
- It creates a system that forces the right things to happen even when those in power have an interest in a divergent outcome.

Fourth Mechanism

- The mechanism attracts the right people and ejects viruses (problems).

Fifth Mechanism

- The mechanism produces an ongoing effect.

(Collins & Porras, 1994)

for the organization and explore how these models accomplished the seemingly impossible (Hall, 2003). This exercise is not about writing empty vision and mission statements; the emphasis is on turning goals into results. Care should be taken to understand what the catalytic mechanisms are that can make a goal reality, providing the link between objectives and performance (Collins, 1999).

In *Built to Last*, Collins and Porras (1994) describe BHAGs as having three key characteristics. First, BHAGs are long term. Leaders should think in terms of becoming the most powerful, most far-reaching, and most successful at what it is their organizations do. They should not sacrifice big long-term goals by focusing on short time frames. Second, BHAGs should be clear and simple. Anything too complicated or nuanced is probably not a large and broad enough goal. Third, BHAGs should connect to the core values of the organization. For example, the BHAG for Granite Rock, a company that sells crushed gravel, is to achieve "total customer satisfaction and a reputation for service" (Collins, 1999, p. 72).

Catalytic mechanisms are the devices to achieve BHAGs. According to Collins and Porras (1994), five characteristics are key to developing such mechanisms, as shown in the box. In essence, BHAGs are an organization's wildest dreams, and catalytic mechanisms are the realistic programs, policies, and initiatives that will allow the organization to achieve those dreams (Collins, 1999). Thinking in terms of leadership, the notion of BHAGs and catalytic mechanisms echoes Level 5 leadership principles—leaders should be able to think big but should remain realistic, focused, and sensitive to input from subordinates. The BHAGs are what police work is all about. The goal of effective policing—policing in a manner that will satisfy both the politicians who deploy police forces and the public who (erroneously) perceive the police as serving them—is a truly big hairy audacious goal.

Police organizations have competing *formal* and *informal goals*. The formal goal of policing is one borne by its outer shield—to serve and protect the public. But simultaneously there is always an informal policing goal—to serve and protect the politician and the political agenda espoused by his/her political reign (for a further discussion of the formal versus informal goals of

Flow	Resonance
Characteristics 1. Time warps, either speeding up or slowing down. 2. One loses self-consciousness. 3. One focuses intently on the task. 4. One performs at the peak of one's abilities. 5. It seems effortless, as if it were flowing. 6. The experience is intensely internally satisfying. 7. Afterward, one regains a sense of a stronger, more capable self.	Resonance is "the sense of seamless harmony with one's surroundings so that internal experience and external experience are one, which produces the fulfillment of performing at your best without strain."
(Csikszentmihalyi, 1990)	(Clawson, 2003, p. 137)

policing, see M. R. Haberfeld, *Critical Issues in Police Training*, 2002). Police organizations must constantly juggle the burden of these competing goals of policing, and they risk making the goal of an effective and functional police organization a pie in the sky.

Bringing this pie in the sky down to earth will be possible only through tiny and incremental steps. It requires using catalytic mechanisms, tools that are effective only when placed in the hands of effective workers. Those workers can be created only through an intensive exposure to leadership training that should and must start on the first day of the basic academy.

Juggling the impossible can actually prove to be impossible, so delivering an illusion is sometimes as good as we can hope for. Police organizations will never be freed from political influences; experiences like the one lived by Chief O'Brien in managing the Elian Gonzales case will always haunt these environments. However, the true challenge is to be mentally and emotionally prepared. Such preparation relies on maintaining a view of integrity that goes beyond one's personal viewpoints to what effective policing should mean to the politician, the public, the chief, and finally the police officers on the street. The notion of compromise of core values and beliefs is not always necessarily negative in nature. Most beliefs about what is right or wrong merely reflect personal perceptions and are not ultimately correct in each and every situation, particularly if one is prepared to look at problems and decision-making processes from the leadership perspective rather than from a self-centered orientation.

Finally, what is the ultimate goal of this book? The ultimate goal is to achieve what Csikszentmihalyi (1990) defines as "flow" or what Clawson (2003) refers to as the "universal sense of resonance" (p. 127). Both authors describe this phenomenon as akin to the emotional sensation felt when things go well. Most people have experienced it (e.g., being "in the zone" while doing sports); individuals can commonly describe a variety of times when they have felt this sense of harmony with their task (e.g., when giving birth to a child or while writing computer code; Clawson, 2003). The goal in the context of leadership is that each one is able to function in a work environment that not only produces the highest level of satisfaction but also enables one to perform at the height of one's ability and, by doing so, to realize the dreams of one's life.

The first requirement in order to get to a point where one can truly achieve this point of flow or resonance is to have a dream—a dream not only of what one wants to be (e.g., a professional basketball player or a chief of police) but of how one feels while one is performing at that pinnacle. Clawson (2003) refers to this experiential dream as an internal dream. To make it come true, however, involves work. This may involve training, practice, study, travel, reading, changing one's lifestyle, or any number of other activities; it also requires maintaining the necessary energy (by removing negative factors and influences) in order to devote all that is necessary to realize the dream (Clawson, 2003). Furthermore, the cyclical process of achieving resonance is inevitably hindered by setbacks, obstacles, and successes, during which time it will be necessary to reinject

Five Key Questions for Resonance Focus

1. How do I want to feel today?
2. What does it take to get that feeling?
3. What keeps me from that feeling?
4. How can I get it back?
5. What am I willing to work for?

(Clawson, 2003)

energy and revisit the dream in order to maintain the vision (Clawson, 2003). The good news is that because the goal is to achieve a state of mind rather than a concrete job or position, it will be possible to catch glimpses of resonance during the preparation process as one comes closer to making that resonance more of a constant reality.

   If there is a work environment that can enable somebody to actually fulfill the dreams of his/her life and at the same time elicit ultimate performance as well as optimum satisfaction and harmony, this environment can and should be found in the field of policing. After all, what more can one hope for in life if not the ultimate sacrifice, devoid of any selfish feelings? Running toward danger instead of away from it—done out of a mission to save others—is the epitome of the flow and resonance concepts, and the epitome of what police work is all about. Finally, to further flash out the concept of running toward danger instead of away from it, regardless of one's gender, a few examples of female police leadership bring the themes touched upon this book into their final accord.

FEMALE POLICE LEADERS—SOME THOUGHTS FOR THE FUTURE

CAREER HIGHLIGHTS

Police Chief Kathy L. Lanier and Discipline

Washington, D.C., Metropolitan Police

Police Chief Kathy L. Lanier believes that paramilitary organizations tend to be too punitive and has adopted a policy that favors training over discipline, stating police departments, in general, have too rigid chain of command and that this is what makes police departments punitive in nature when it comes to discipline.

   She tries to make a distinction between deliberate rules violations and simple mistakes or violations that resulted from an officer(s) not understanding the meaning of a law or policy. For the ladder, Lanier says this "if I discipline you, are you going to understand it any better?" (Cella, 2011, p. 1). If the officer violated the policy because he/she did not understand the policy, punishing him/her punitively is ineffective since the punishment does not seek to address the root cause of the rule violation, ignorance, and instead, simply seeks to punish the individual for his/her misdeed. According to Lanier, in these types of situations, it is more beneficial to send the officer for additional training so that he/she can come to understand or garner a better understanding of the policy that they violated so that he/she will not violate that policy again in the future (Cella, 2011).

National Park Police Chief Teresa Chambers and Politics

Police Chief Teresa Chambers was the first woman to ever become the Chief of the Park Police and the first Park Police Chief to come from outside of the department (Barker, 2004).

On December 5, 2003, Park Police Chief Teresa Chambers was suspended just days after making complaints that her department was underfunded and understaffed ("Suspension," 2003). On December 12, 2003, U.S. Park Service offers to reinstate Chambers if she agrees to have her press releases reviewed by her boss; Chambers refused the offer ("U.S.," 2005). Her attorneys filed a petition to immediately reinstate her as Chief with the Merit Systems Protection Board. Later that day, she was fired. On July 10, 2004, Chambers accused the Bush Administration of "silencing views in the rank and file" ("Ex-Chief," 2004, p. 1).

Chambers was not discouraged by this turn of events and on January 10, 2008, became the Chief of the Riverdale Park Police Department (Prabhu, 2008). On January 18, 2011, Chambers announced that she will be returning to the U.S. Park Service as Chief of the Park Police. On January 30, 2011, Chief Chambers resigned from the Riverdale Park Police Department and resumed her job as Chief of the Park Police (Dasgupta, 2011a, b; McCartney, 2011).

The Glass Ceiling theory of female leadership asserts that women lack political savvy (Northouse, 2003). Chambers cannot be accused of lacking political savvy; although her spin on politics can be certainly viewed as "different" to say the least. She stood up by her principles and subsequently lost her job; nevertheless, she took on the federal government and the President of the United States when she was fired from the U.S. Park Police—a controversial, yet a very bold, move that would ultimately result in getting her job back years later.

Commissioner Kathleen O'Toole and Reforms

Boston Police Department, Inspector General of the Irish National Police Force

Commissioner O'Toole obtained her Bachelor of Arts from Boston College and Juris Doctorate from New England School of Law (Kiernan Group

Holdings, Inc., n.d.). Prior to becoming the Commissioner of the BPD, she had a long and successful career in public and private sectors.

On February 8, 2004, Kathleen O'Toole was appointed Commissioner of the Boston Police Department. Soon after her appointment, on July 19, 2004, O'Toole announced reforms in evidence collection policies and policies pertaining to the construction of photo lineups to help decrease wrongful convictions. Following her commitment to overhaul the department, on October 13, 2004, O'Toole shuts down the department's fingerprint unit to revamp it because of past troubles within the unit. In an attempt to introduce new accountability mechanism, in December 2004, O'Toole tries and fails to start a civilian review board to investigate complaints against BPD officers. A year later Boston reports 75 homicides in 2005, a 10-year high; clearance rate for homicide also falls below 33%. On May 9, 2006, O'Toole resigns from BPD to become the Inspector General of the Irish National Police Force (Pratt, 2006; "Timeline," 2006).

The event that defines O'Toole's leadership was the death of Victoria Snelgrove. In October of 2004, Victoria Snelgrove was shot in the eye by a pepper spray projectile. The shooting occurred during a riot following the Red Sox's advancement to the World Series over their rival the New York Yankees. According to eyewitness, the death of Snelgrove was caused by an unknown rioter throwing a bottle at a mounted police officer. The bottle landed near the officer, startling his horse. In response, a nearby officer turned and fired into the crowd, in the direction that the bottle came from, striking numerous people. Snelgrove died of her injuries just hours after being shot. Following an investigation, O'Toole placed two officers on leave as a result of the incident. On October 22, 2004, O'Toole announces that the Boston Police Department accepts full responsibility for Snelgrove's death. However, O'Toole also publically condemned the rioters and stated that they also "own a portion of the responsibility for Snelgrove's death" (Farragher & Abel, 2004; Police, 2004; Sukiennik, 2004). The follow-up investigations conducted by the Internal Affairs confirmed that the officers made a host of serious errors during the incident but that those errors represented poor judgment and not criminal intent; subsequently the officers faced no charges since their poor judgment did not warrant being charged criminally (The Washington Post, 2005; Zezima, 2005).

According to some researchers, females tend to be evaluated unfavorably when they use a directive or autocratic style (Dreher & Cox, 1996; Eagley et al., 1992). When O'Toole declared that the five officers would not be charged in Snelgrove's death, she was sharply criticized by the Snelgrove family. Furthermore, when she participated in the Independent Commission on Policing in Ireland (directive in nature), the findings of the commission were bashed by many. Morris (1998) conjectures that women lack self-confidence which presents much of a challenge in controversial police environments. However, O'Toole attended the funeral of Victoria Snelgrove, despite the accusations that misconduct on the part of her officers played a part in her death, displaying a healthy dose of self-confidence given the nature of the charges. Similarly, she was not afraid to ask for additional resources or permission to make sweeping changes in policing tactics. O'Toole never seemed to be swayed by criticism of her actions.

Krieger (2003) posits that retention of talented women is an issue. Kathleen O'Toole is without a doubt a talented woman, but her resume shows than she has not stayed with one organization for a long period of time. Was it based on her desire to expand in different directions or was it a sign of her disappointment with lack of adequate support provided from within the organization for a strong leader who is not afraid to innovate and make statements however, at the same time, requires more organizational support due to the gender stereotypes?

Politics, discipline, and reforms represent some of the leadership challenges that women were, historically, assumed to be less capable of handling effectively than their male counterparts. Yet, the three female police leaders featured in this chapter prove that female police leaders can be as skillful as males when tackling these challenges.

Time has come in the 21st century, and it is long overdue, to overcome this "Big Hairy Audacious" obstacle of stereotyping females as less capable of handling high ranking positions and include more women in top police leadership ranks making equality and parity a truly meaningful phrase.

References

Barker, K. (2004, July 10). Park police chief fired after dispute, suspension. *The Washington Post*. Retrieved June 7, 2011, from http://www.washingtonpost.com/wp-dyn/articles/A39354-2004Jul9.html.

Bryant, A. (2003, June 23). Breakout best seller. *Newsweek*, 141 (25), 50.1.

Cella, M. (2011, June 2). Lanier elaborates on police discipline complaints. *The Washington Times*. Retrieved June 7, 2011, from http://www.washingtontimes.com/blog/city-state/2011/jun/2/lanier-elaborates-police-discipline-complaints/.

Chief of U.S. park police fired. (2004, July 10). *Milwaukee Journal Sentinel*. Retrieved June 7, 2011, from http://news.google.com/newspapers?id=AbcaAAAAIBAJ&sjid=REUEAAAAIBAJ&pg=2468,7313272&dq=teresa+chambers&hl=en.

Clawson, J. (2003). *Level three leadership: Getting below the surface*. Upper Saddle River, NJ: Pearson Education, Inc.

Collins, J. (1999, July–August). Turning goals into results: The power of catalytic mechanisms. *Harvard Business Review*, pp. 71–82.

_____ (2001). *Good to great*. New York: Harper-Collins.

_____ & J. Porras (1994). *Built to last*. New York: Harper-Collins.

Csikszentmihalyi, M. (1990). *Flow: The psychology of optimal experience*. New York: Harper & Row.

Dasgupta, S. (2011a, January 18). Chambers imminent departure would mean search for new police chief. *Riverdale Park-University Park Patch*. Retrieved June 7, 2011, from http://riverdalepark.patch.com/articles/chambers-imminent-departure-would-mean-search-for-new-police-chief.

Dasgupta, S. (2011b, January 24). Riverdale Park Police Chief takes her former position with U.S. Park Service. *Riverdale Park-University Park Patch*. Retrieved June 7, 2011, from http://riverdalepark.patch.com/articles/riverdale-park-police-chief-takes-her-former-position-with-us-park-police.

Dreher, G. & T. Cox (1996). Race, gender and opportunity: A study of compensation attainment and the establishment of mentoring relationships. *Journal of Applied Psychology*, 81, 297–308.

Eagley, A., M. Makhijani & B. Klonsky (1992). Gender and the evaluation of leaders: A meta-analysis. *Psychological Bulletin*, 117, 125–145.

Ex-chief of park police denounces firing. (2004, July 10). *CNN*. Retrieved June 7, 2011, from http://articles.cnn.com/2004-07-10/us/park.police.chief_1_national-park-service-dissenting-views-duties?_s=PM:US.

Farragher, T., & Abel, D. (2004, October 22). Postgame police projectile kills an Emerson student. *The Boston Globe*. Retrieved June 13, 2011, from http://www.boston.com/sports/baseball/redsox/articles/2004/10/22/postgame_police_projectile_kills_an_emerson_student/.

Hall, B. (2003, November). BHAGs for 2004. *Training*, 40 (10), 16.

Kiernan Group Holdings, Inc. (n.d.). *Advisory board-Kathleen O'Toole*. Retrieved from Kiernan Group Holdings, Inc. website: http://www.kiernangroupholdings.com/index.php?option=com_content&view=article&id=49&Itemid=41&limitstart=7.

Krieger, J. (2003). "Women and Leadership" in Leadership: Theory and Practice (3rd ed.). Thousand Oaks: Sage.

Lenzner, R. (2003, April 28). Room at the top. *Forbes*, 171 (9), 68–71.

Macht, J. (2001, October). Jim Collins to CEOs: Lose the charisma. *Business 2.0*, 2 (8), 121–122.

McCartney, R. (2011, January 21). Park Police Chief's Saga a triumph. *The Miami Herald*. Retrieved June 7, 2011, from http://www.miamiherald.com/2011/01/21/2027748/park-police-chiefs-saga-a-triumph. html#storylink=misearch.

Morris, B. (1998, October, 12). The trailblazers: Women of Harvard's MBA Class of '73. Fortune, pp. 106–125.

Northouse, P.G. (2003). Leadership: *Theory and Practice* (3rd ed.). Thousand Oaks: Sage

Pickard, J. (2003, November 6). Collins challenges charm culture. *People Management*, 8.

Pratt, M. (2006, May 9). Boston police chief quits for Ireland job. *The Washington Post*. Retrieved June 13, 2011, from http://www.washingtonpost.com/wp-dyn/content/article/2006/05/09/AR2006050901363.html.

Prabhu, M.T. (2008, January 17). Riverdale Park gets new chief. *Business Gazette*. Retrieved June 7, 2011, from http://www.gazette.net/stories/011708/landnew184032_32365.shtml

Sukiennik, G. (2004, October 23). Boston fan dead after police fight with rowdy crowd. *Portsmouth Daily Times*. Retrieved June 13, 2011, from http://news.google.com/newspapers?id=SzxGAAAAIBAJ&sjid=NNEMAAAAIBAJ&pg=5626,4096399&dq=victoria+snelgrove&hl=en.

Suspension given to police chief. (2003, December 6). *Lakeland Ledger*. Retrieved June 7, 2011, from http://news.google.com/newspapers?id=19NOAAAAIBAJ&sjid=3v0DAAAAIBAJ&pg=4266,2782418&dq=teresa+chambers&hl=en.

Timeline of Kathleen O'Toole's tenure as police commissioner. (2006). *The Boston Globe*. Retrieved June 13, 2011, from http://www.boston.com/news/local/massachusetts/articles/2006/05/09/otoole_timeline/.

The Washington Post. No charges against police who killed Red Sox fan. Retrieved June 13, 2011, from http://www.washingtonpost.com/wp-dyn/content/article/2005/09/12/AR2005091202118.html.

U.S. Park Service offered to restore Chambers as chief. (2005, March 31). *WTOP*. Retrieved June 6, 2011, from http://www.wtop.com/?sid=162193&nid=25.

Zezima, K. (2005, May 26). Panel blames police for errors in fan's death. *The New York Times*. Retrieved June 13, 2011, from http://www.nytimes.com/2005/05/26/national/26fan.html.

In Their Own Words

Over 18,000 police agencies in the United States represent over 18,000 different police leadership styles times as many officers as these agencies employ. It is beyond the scope of any book to present their thoughts and insights into various events that required some sort of leadership decision. It is only natural to finish this book with some approaches into leadership behaviors accounted for by the leaders themselves rather than by this author. The few cases presented below illustrate certain situations depicted in the unedited words of a number of police leaders, as the title of this chapter hints to, and the readers are challenged to match the leadership theory that best fits their behaviors.

Chief John DeCarlo
Branford, Connecticut Police Department

OPEN COMMUNICATION LEADERSHIP

During the spring of 2010, a brutal murder of Yale Medical School postdoc, Vajinder Toor, took place in the parking lot of a condominium complex in Branford, Connecticut. A former coworker, Lishan Wang, gunned down Toor as he walked from the front door of his condo to his automobile which was parked yards away in a common parking area. Wang secreted his pistol in a paper bag and approached Toor closely before he shot him multiple times.

In addition to mortally wounding Toor from close range, Wang also opened fire on Toor's pregnant wife who was waving goodbye to her husband from their porch as the family's three-year-old waited inside the condo. Toor's wife was not hit and was able to retreat to the inside of their home. The murder took place in the northern part of Branford that is densely populated by residents living in close proximity in well-maintained, upscale condominiums. At the time of the murder, people were going to work and children were on their way to school. As is the case in many residential neighborhoods, people were jogging and going about other routine activities in proximity to where the murder took place.

Our police department was notified by telephone that there had been shots fired and a person was seen leaving the scene in a van. Responding police officers spotted the van, containing Wang, making his way to Interstate-95, a short distance south of where the murder had taken place. The van was stopped and Wang was taken into custody within minutes after he left the scene of the crime. Obviously, many police cars and soon after media vehicles were at or near the scene of the crime. While I was present at the scene, I realized that the fear level of the many residents of the area was extremely high and people were understandably reticent to leave their homes.

Recalling George Kelling's article, Crime and Metaphor, I knew that keeping the public in the loop was vital and something often overlooked by police departments. I felt that since the investigation was well established that the police department should take the initiative in letting the community know that the danger was over and that their families could safely go about their lives. I instructed our public information officer to immediately advise the press that the suspect was in custody and that the danger was over. In addition, I instructed police staff to issue a reverse-911 telephone message to all of the homes and businesses for about a mile around the crime scene, that the police had taken the perpetrator into custody and that they were not in any danger.

Bad news and rumor travel quickly and can cause fear in a community. When police are in front of the information curve and can share facts of a crime with residents, it helps alleviate fear of crime and builds confidence in and feeling of partnership with the police. Our agency received an unprecedented number of calls and letters after this case thanking the department for quickly letting them know that they were safe and that the danger was over. Sometimes leadership is as simple as a phone call.

Deputy Chief William Fraher
Paterson Police Department, NJ

EQUITY LEADERSHIP

In the late 1990s, I was assigned command of my department's communications division after an anonymous letter to the local newspaper alleged serious acts of misconduct by some of the personnel assigned. The chief complaint was that certain supervisors were absenting themselves from the center during the tour of duty and that they were falsifying official documents intended to control attendance. The department's Internal Affairs unit and the local county prosecutor initiated an investigation and my job was to fix the problems.

The division was composed of civilians and sworn personnel that worked in three shifts. I found that inattentive and apathetic leadership had allowed factions to develop on each shift. The "in-group" members were given priority in just about every aspect of the division's operations. Among the complaints were priority for overtime assignments in violation of contractual provisions, days off, duty assignments, lunch assignments (the unit had a practice of "early" and "late" lunches that allowed staff members to come in late or go home early), and tour assignments. Morale was low and there was constant bickering among the staff.

Equity issues were one of the more common complaints. In the staff's view of the situation, rewards were not being administered fairly. They assessed their inputs to outcomes ratios and found them to be out of balance. As a result, most staff members were adopting resolution strategies that threatened to compromise the unit's effectiveness.

To restore the equity of the individuals, steps were taken to assure that all issues concerning the employment contract were handled according to the terms of the contract. Seniority rights were to be respected when it came to tour assignments, day off and vacation requests, and overtime. All other rewards were to be tied to actual work performance. A reporting system was put in place to ensure that all required time and attendance records were completed in a timely fashion. Several other management processes were initiated to ensure compliance. At a meeting with the subordinate leaders and tour supervisors, they were told in no uncertain terms what was expected of them. Failure to follow any of the department policies, rules and regulations would be considered a serious breach of discipline. To further reinforce the seriousness of the situation, inspections of the unit were carried out at random times to check for compliance; everyone was kept on a short leash.

In the short term, I was not the most popular guy with some staff. They were used to absentee leadership. They did not like the extra scrutiny, and I didn't care. The communications function of any police department is critical for the citizens and police officers that depend on them. Anything that affects performance for the worse can have serious consequences. Over time things improved, not everyone was happy but they recognized that they were being treated fairly.

James R. Davis
Chief of Police (Ret.)
El Cajon Police Department
El Cajon, California

EMOTIONAL EQUITY LEADERSHIP
THE FIRST INVESTMENT FOR SUCCESS

Of all the things I count as accomplishments during my first 30 years in public safety, one of the most satisfying projects was the passage of Proposition "O." Passed on the November 2004 ballot, Proposition "O" was a measure to increase the sales tax within the City of El Cajon. The money was to be earmarked for the construction of a new public safety facility. The term "public safety" was used deliberately. My vision for the facility was to incorporate police and fire operations as well as related community services. The funding would also support the renovation of two existing fire stations and the animal rescue shelter.

This project had its origins in 1974. Our Chief of Police told us we were three to five years away from moving into a new police station. Three successive chiefs would echo the same message. Many things contributed to the delays. No matter the reasons, the result was the same. In a city of only 14.2 square miles, the department was forced to decentralize into three old, overcrowded, and inadequate facilities. This forced decentralization was bad for communication, efficiency, and morale.

As the Chief of Police in 2004, I was determined to find a way to bring together the collective time and talents of those in the community to provide a public safety facility worthy of the community it would serve. In 2004, after discussing the options with the City Manager and the Assistant City Manager, we initiated a process to explore how we would fund a new public safety facility. From the beginning, it was our commitment that this would be a collaborative process involving all our stakeholders. Our team examined a variety of funding options and spoke to city and community leaders throughout the state about their experiences. I recognized that we needed to push beyond the self-imposed boundaries that had thwarted prior attempts. With the help of a specific area expert, we concluded that our best and most cost-effective option would be to ask the citizens of El Cajon to increase their sales tax by one half cent.

We faced many challenges. Few gave us any chance of being successful. Tax increases for defined purposes require a two-thirds majority. No one in our region had an additional half cent sales tax. Tax increases in San Diego County, and more specifically East County, were rarely successful. Many major businesses within the city, although supportive of the police and fire departments, were concerned about the effect that an increased sales tax would have on their business. The support from some elected officials was less than enthusiastic. Even our specific area experts gave us less than a 50–50 chance of success.

In the campaign for Proposition "O," I put forth a vision not only of need but of a different style of service delivery; a style that included greater sharing of resources, more effective communication, more inclusion of community, and greater access to community resources. Volunteers from the department, retired officers, the community, and I invested the time to explain the need and share the vision with stakeholders. Reactions were mixed. Many business owners were concerned about the impact on sales. Some saw the need and agreed to help while others recognized the need but objected to any new taxes. Almost without exception, people appreciated the candor and clarity of the discussions.

With the support of a strong City Manager, and an equally tenacious Assistant City Manager, we made the decision to move forward. We got the right people in the right positions. This concept would prove to be critical to our victory. Through a strong collaborative effort, using the skills of specific area experts regardless of their position within the organization, we developed a sound plan and an honest and clear message. With a sound plan and clear message in place, we relentlessly worked the plan and delivered a consistent message. Through a series of frequent regular meetings, discussions, disagreements, resolution of disagreements, and decisions on how to move forward, we constantly made appropriate adjustments to the plan while continuing to display the discipline to stay on message.

On election night, November 2, 2004, the voters of El Cajon gave the measure the required two-thirds majority and the measure was passed.

Many experts were surprised by our victory. It is impossible to overemphasize the importance of the sacrifice of all our volunteers. Without their constant dedication, time, and talent, the victory would not have been possible. I am, however, also convinced our margin of victory was based, in no small part, on something I term "Emotional Equity."

Years of dealing with challenges and crisis in a transparent manner, of being active, involved, and invested in the community (I don't make a distinction between being on and off duty. Police Command Officers, and arguably all police officers, are never really "off duty," but that is another discussion) of including the community in import decisions about the staffing and direction of their department. Decisions such as officer selection, promotions, and service priorities are a few that saw community input if not direct participation. We always worked for the goal of "do the right thing." I have no doubt that the way our officers and staff interacted with the community resulted in a strong reserve of Emotional Equity. While the community may have been nervous about the unforeseen consequences, they liked the vision and they trusted us to do the right thing.

James R. Davis
Chief of Police (Ret.)
El Cajon Police Department
El Cajon, California

SEPTEMBER 11, 2001—PREPARATION LEADERSHIP

There is no substitute for building strong ties with the community and nurturing personal relationships with the people we serve. The bonds of mutual trust and respect fill the Emotional Equity accounts that are drawn upon in times of crisis.

Tuesday morning I was up early. My wife was away visiting relatives. I wanted to get an early start on the day. As is my habit, I was listening to the news as I shaved. My regular morning program was interrupted by news of an explosion in the North Tower of the World Trade Center. Although this seemed to be a potentially tragic incident, it did not seem to be anything that was likely to have an immediate effect on my community of El Cajon in Southern California.

Over the last 20 years El Cajon has become highly diverse. Evolving from a mostly white English speaking community, there are approximately 50 different primary languages spoken by the students in the middle school. We are home to the offices of the Kurdish Human Rights Watch and the third largest population of Chaldeans (Christian Iraqis) in the world. At that time, one of three mosques in San Diego County was located within the City with two more lying less than 5 miles from our borders. All the independent markets and many other businesses are owned by people who recently emigrated from the Middle East or surrounding regions. We are also home to the club house for the San Diego Chapter of the Hells Angels and, as part of Eastern San Diego County, have a reputation for our share of less than tolerant individuals.

Between 8:46 AM and 10:03 AM, four planes were hijacked, three had been used as weapons against the Twin Towers of the World Trade Center, and the Pentagon and the fourth had crashed into a field in Pennsylvania. Descriptions of the people who commandeered the planes changed from hijackers to terrorists. The face of the terrorists would look a lot like many of the people in my community, a Middle Eastern male. This was clearly going to affect my community.

I contacted my communications center and asked them to call my Captains for an early meeting. As we assemble that morning, I had three, inextricably linked, priorities: 1) to ensure, to the extent possible, our officers had the information they needed to do their jobs effectively and safely; 2) to prevent what I thought might be a backlash against our citizens of Middle Eastern descent; and 3) to find ways to get information to all segments of our community and by doing so provide for the safety and security of all our citizens.

Over the ensuing days and weeks we implemented a variety of strategies to accomplish these goals. We kept our officers and professional staff formed. Officers were asked to make a point to stop and talk with business owners and individuals in their areas of responsibility and assure them that we were there to protect them and to encourage them to report all crimes (including hate crimes) or suspicious activity. As a tangible display of solidarity with their new country, we encouraged them to display American flags outside their businesses. We organized community meetings featuring speakers who were Pastors of Christian churches, the Imam from a local mosque and other community leaders. We met with formal and informal community leaders from the immigrant communities and encouraged them to find ways to communicate the depth of their loyalty and commitment to this country. Blood drives were organized at the Kurdish Human Rights Watch office and patriotic rallies were among the expressions of solidarity. American flags were everywhere.

Ultimately we exceeded the original goals. Thanks to preexisting strong lines of communication between regional public safety agencies, our officers and professional staffs received updates as they became available. While there is, of course, no way to empirically prove this, I believe that our efforts to communicate with our community and encourage cross-cultural communication along with high-visibility directed patrol resulted in zero reports of anything approaching a hate crime.

I'm convinced that there are three themes prevalent in our success. All three are linked and without one the other two would not have been as effective. They are **Being Proactive, Communication,** and **Preparation**.

While no one could know what the effects would be of the events of that September morning would be, our understanding of our community, prior experience, and understanding the lessons learned by other departments served to inform our actions. We knew our community's strengths and weaknesses. Based on the nature of the events we were able to anticipate, at least the possibility if not the probability, of certain types of crime and form plans that simply built upon our existing philosophy of Community Based Policing. Because of years of establishing mutual trust and confidence within the community we were able to get out ahead of the rumors and misinformation, which spawn fear, with a clear and consistent message. Positional and non-positional leaders from the community had long been included in my informal advisory group. I was on a first-name basis with all of them. I had their direct phone numbers and they had mine.

As they demonstrated their connection to us, so the department demonstrated its connection to them. Our officers and professional staff took every opportunity to listen to and keep our citizens informed. My staff and I were seldom out of uniform for several weeks after September 11. I personally made a number of contacts with leaders within the Middle Eastern community and the community at large. I was present, in uniform, at every public event. When the Kurdish human rights watch office held a blood drive, I was there to donate blood.

Drawing upon a reserve of Emotional Equity in the community we were able to bring diverse groups of people together to listen to faith and secular leaders give voice to their thoughts and concerns. Many in the immigrant community, who were at first fearful, accepted our suggestions and chose to be, themselves, proactive and find ways to demonstrate their commitment, love, and appreciation for our country. Through patriotic rallies, blood drives and other events designed to demonstrate solidarity, the Middle Eastern community was able to show tangible evidence of their opposition to those who had attacked us.

This level of Emotional Equity is not established over night. It is the result of years of building on the work of those who came before us. Refining and, where necessary, reshaping the organization to meet the demands of change. We were second only to the San Diego Police Department to establish Community Policing. While, like everyone else, we had difficulty finding just exactly the formulas for success (I prefer to think of it as an evolutionary process), we continued to find ways to mold our philosophies of service delivery to evolving community needs and expectations.

Setting aside for a moment the immense personal and professional satisfaction I believe is associated with the opportunity afforded to public safety professionals, to meet with and get to know the people who are affiliated with the myriad of individual special interest, civic, benevolent, religious and all types of civic groups in a community; it was those established relationships along with our reserve of Emotional Equity that allowed them to trust our leadership in this time of crisis.

The mission statement of the El Cajon Police Department reads: "Committed to a Safe and Secure Community through Service, Mutual Cooperation and Respect." While I understand that one can't prove a negative, the City took pride in the absence of even a single report of a hate crime. The Department, by always striving to live up to the values that underpin the mission statement, had established the level of Emotional Equity that paved the way for the preparation, communication, and proactive efforts to be effective.

James R. Davis
Chief of Police (Ret.)
El Cajon Police Department
El Cajon, California

CRITICAL INCIDENT LEADERSHIP

The anguish and fear was visible on the faces of the parents and relatives of the 2,800 high school students, their teachers and staff members evacuating the high school in the solidly middle-class neighborhood. Just before 1 PM shots and rang out and now, for the second time in 17 days and less than 6 miles away from the first incident, the high school in the Grossmont Union High School District was under attack by a teenage gunman.

The shooter, an 18-year-old student, purchased a 22 caliber pistol and took a shotgun which he loaded with bird shot in preparation for his shooting rampage. He would later tell a probation officer he "wanted to do suicide by cop."

At approximately 12:54 PM, Jason Hoffman drove his truck to the bus loading area of Granite Hills high school, got out and, with his shotgun in hand then 22 caliber pistol in his belt, walked quickly on to the school campus immediately confronting high school Dean, Dan Barnes (Dean Barnes had counseled Hoffman just days prior regarding disciplinary issues. Dean Barnes would recall that the meeting seemed to have ended on a positive note) and started shooting. The shooting lasted about 90 second. That minute and a half resulted in one staff member and four wounded students and school and community rocked by, yet another, school shooting. If not for the actions of the School Resource Officer (SRO) Richard Agundez and San Diego Sheriff's Deputy Angela Pearl, the results would have undoubtedly been quite different. The officer and deputy were just steps away at a small party in the administrative area. Both ran to the sound of gunfire.

The first six 911 calls were all from students and staff with cell phones. Arriving officers quickly became aware that the primary suspect was in custody, shot and wounded by Officer of Agundez. The potential existence of another suspect was an open question.

Acting on a new tactical paradigm, SWAT officers did not wait for all of their team members and equipment to arrive. With minimal equipment, some still in shorts, the ad hoc team of SWAT officers from our agency and the Sheriff's office formed up and began a sweep of the campus. Students and staff followed previously established instructions and, with the aid of other police officers, evacuated to a nearby park to be reunited with the rapidly growing number of relatives and parents. The geography of the area worked in our favor. With the aid of short fences from ballparks and the sheer size of the area we were able to establish distinct areas for student evacuation, assembly of parents and relatives, officer assembly area, command post, and appropriate media assembly area. Thanks in no small part to the recent experience with at Santana High School, good planning and preparation by patrol commanders and experienced senior sergeants on scene helped the law enforcement function extraordinarily well.

The event—coupled with the shooting less than three weeks earlier at Santana, a high school in the same district—served to make this a major news event. The streets leading to the school were soon lined by news vans and a forest of satellite antennas. The events of that day were beamed around the world. It is of course difficult to judge the reach of events like this. I am sure it made people, who never heard of Granite Hills High School before, feel more vulnerable. It certainly made the community feel that way.

This sense of vulnerability was not lost on key leaders. The Grossmont Union High School District Superintendent, San Diego County District Attorney, San Diego County Sheriff, Superintendent for the Cajon Valley Union School District and I met and, in an atmosphere of cooperation, quickly formed a plan to work with the press and communicate with our citizenry. This process was made much easier by the fact that we have all previously made a point to work together and by doing so had formed a high level of mutual respect. We quickly agreed on a message that we wanted to convey to not only our community but also other communities that might be affected by this incident. The message was simple and would form the foundation of the myriad of interviews that would follow. Through the media we conveyed our message to the community that schools were still among the safest places for young people to be.

In a series of frequent and regularly scheduled press conferences, we appeared together, we told the truth (like any good interviewer the press often asked questions for which they already have what they believe to be an answer), as long as it did not jeopardize the integrity of the investigation, and we gave them the details regarding the investigation and were sensitive to their deadlines. We made trained public information officers available to them to facilitate logistical issues. Key leaders appeared at each press conference. We answered questions candidly and always returned to our basic message. Dozens of calls were received requesting for me to appear on a variety of national and international radio and television programs. I resisted the lure to stroke my own ego by appearing on all that ask. The constraints of time and good judgment dictated that I prioritized my appearances to those that I thought would convey our message most effectively and benefit my community most.

After the first school shooting, at Santana High School, we held some public meetings to help gauge the public concern and answer any questions. Ironically one of the first meetings was held in the gymnasium at Granite Hills High School. It was very sparsely attended.

After the second school shooting, the community meetings, including one at Granite Hills High School, were all standing room only. All of the key leaders attended and fielded questions on topics ranging from the law, prosecution of the offender, school security in the form of fences and metal detectors, and the efficacy of stationing police officers at each school. We all stayed as long as was necessary to answer all the questions.

I believe our individual and collective reputations, established levels of Emotional Equity, experience, responsiveness to their concerns, and unified message allowed school to reopen quickly with a better than 93% attendance rate.

We reassured the community of our commitment to safe schools, thorough investigations, appropriate prosecutions and a transparent review of the issues. Officers were assigned to the Middle Schools and High Schools for a period of time. We worked closely with our schools and surrounding agencies to further refine school staff and law enforcement responsibilities during critical incidents on campuses. Drawing on a field of specific area experts, academics, and community leaders, the Superintendent of the GUHSD established a Lessons Commission of which I was privileged to be a part. After extensive study of the issues, input from school faculty and staff, students, and parents, the commission issued a report including a series of recommendations that were unanimously adopted by the School Board.

The evening after the shooting, I visited all of the victim and their families. Frightened, angry, and shaken by the events of the day they seemed appreciative of the response. I am proud of how our Department acted. The margin of success was a combination of the right people in the right places, preparation (Emotional Equity), a sound plan, discipline in the execution of the plan, and a tireless commitment to those we serve.

Chief Robert F. Vodde
Police Chief (Ret.)
Leonia Police Department
Leonia, New Jersey

SITUATIONAL LEADERSHIP

Effective leadership entails innumerable traits and qualities, not the least of which include problem solving, decision making, and an adherence to strong ethical and professional principles. While situations involving legal, procedural, and operational decisions are often associated with one's leadership style and ability, matters involving personnel are less glamorous and desirable, often because they can adversely impact morale, despite the importance for maintaining structure, discipline, and the integrity of the organization.

Dr. Robert F. Vodde, Director of the School of Criminal Justice and Legal Studies at Fairleigh Dickinson University, who studied leadership while at the FBI National Academy, as well as having taught classes on the subject, explains that effective leadership entails innumerable traits and qualities, not the least of which include problem solving, decision making, and an adherence to strong ethical and professional principles. He explains that while situations involving tactical and operational decisions are often associated with one's leadership style and ability, matters involving personnel are less glamorous and often more difficult, often because they can adversely impact morale, despite the importance for maintaining structure, discipline, and the integrity of the organization.

Addressing one of his own experiences, he writes, "Before entering academia, there was one particular situation that helped define my leadership while serving as Chief of Police in the suburban community of Leonia, New Jersey, located minutes outside of New York City. A few years after becoming Chief, one of my former patrol partners was alleged to have committed several criminal offenses that were related to a substance abuse problem."

While it appeared common knowledge that the officer had a "drinking problem" which would occasionally lead to verbal and physical altercations, and the fact that he was a volunteer firefighter, a long-time resident, and generally considered a "nice guy," made those close to him turn a blind-eye to his indiscretions associated with alcohol abuse.

On one particular occasion while off-duty and drinking, he was alleged to have assaulted a fellow firefighter, which prompted the victim's insistence on signing a criminal complaint against the officer. "By the time I arrived at police headquarters the next day, it was the talk of the town," the Chief explained. As a result of the firefighter's complaint, "I initiated the requisite internal affairs investigation which some of the "old-school" officers believed should have been squashed." After a thorough review of the facts and circumstances, which included extensive eyewitness accounts, it was determined that the incident did in fact occur and was related to the officer's inebriation. Independent of the findings of the internal affairs investigation, the officer was found guilty in criminal court of simple assault, which prompted calls for his dismissal among some other members of the department, elected officials, and the community.

Notwithstanding the implications of the officer's actions on his own reputation and that of the department and community, "I was of the opinion that it was important to keep an objective perspective and take all things into consideration, including his otherwise positive contributions to the department and community, despite his indiscriminate use of alcohol. Despite pressure for a harsh and punitive disciplinary approach, to include his dismissal, I decided that a progressive disciplinary approach would best serve the interests of the officer, the department, and the community. As part of my recommendations to address the officer's problem and associated behavior, I required the officer to participate in an intensive alcohol rehabilitation program as a condition of continued employment, the cost of which would be covered by the department."

"Fortunately," Vodde explained, "the rehabilitation program proved successful in addressing the problem of alcohol abuse, transitioning the officer back to duty, and reconciling his differences with those he offended, despite his ill-feelings over my decision." While retrospectively, the appropriate course of action may appear to have been fundamental at the time of the incident, it was highly charged and controversial, one that challenged my leadership as a new and one of the youngest Chiefs in the State. "I took many factors into consideration in deciding the appropriate course of action," the Chief explained, which included "the officer's legal rights; his past performance; his substance abuse problem and associated behaviors; the victim's need for justice; the morale, discipline and continued smooth operation of the force; and lastly, the reputation and integrity of the department, and community's confidence in its ability to function as a professional organization."

Chief Vincent Del Castillo (Ret.)
N.Y.C. Transit Police Department

ACCOUNTABILITY LEADERSHIP

For background purposes, civilian complaints are allegations of police misconduct made by a citizen who was the subject of the misconduct or a witness to the event. After investigation, those complaints are either substantiated or unfounded (or unsubstantiated). Substantiated complaints are usually handled disciplinarily and if proven true may result in a reprimand, suspension from duty or dismissal from the force.

Shortly after taking command of the 3,500 officers from Transit Police Department in early 1987, I discovered that the rate of civilian complaints against police officers had been increasing steadily reaching slightly over 100 complaints per month. I immediately called a meeting of my top commanders to discuss strategies and options. One of the decisions we had to make was whether or not to go public with the problem. I decided that it would be better to get out in front of the issue rather than have it leak to the press thereby putting us on the defensive. A public meeting of the Transit Authority was scheduled within the next several days. At those meetings, each department gave a status report and discussed current topics with the Transit Authority President and with members of the M. T. A. board. I decided to raise the issue of civilian complaints at that time. It was also important and expected to show what the Department intended to do in response to the problem.

Following the problem-solving model, we first needed to understand the nature and scope of the problem. I appointed a small group of commanders to review and analyze all the complaints received within the past year. Of particular interest were similarities among those complaints, trends, the number of officers involved (*were a few officers receiving many complaints or was it a wide-spread problem*), and finally a breakdown of the typical scenarios and categories in which those complaints fell. The findings of that review helped shape our response. We embarked on a comprehensive plan of action that included training, administrative measures aimed at streamlining reporting and investigative procedures, disciplinary actions, and changes in deployment strategies.

Following are some of the more notable findings and our responses:

- Relatively few patrol officers received large numbers of complaints, many officers received one or two complaints for the year, and the majority of officers did not receive any complaints. We found little correlation between the activity (summonses and arrests) of officers and the number of complaints received. Most of the scenarios read like a script from a TV show or police movie like Dirty Harry or Lethal Weapon where the police would grab or shove someone who had committed a minor offense and curse him out prior to giving them a summons. Apparently, television and movies were the major influence on their perception of the police role and behavior. We attributed this problem to the hundreds of new officers hired within the past four years. Many of the more recent officers received their field training from officers who themselves had only a short time on patrol.

  In response, we developed two separate in-service sensitivity training programs: a two-day program for officers who received one or two complaints and a one-week program for officers with several complaints. These programs emphasized human dignity, civility and the impact of police service on the public's image of police. We believed that the officers were sincerely trying to fulfill their role but lacked the necessary skills.

- A large number of complaints resulted from school condition assignments where officers were randomly assigned to stations adjacent to public schools around dismissal time. Students were often raucous and noisy with a great deal of horsing around and other minor mischief. Some officers were apparently overwhelmed with the disorder and responded with unnecessarily roughness in their attempts to control those students.

  We came to realize that with police assignments, police are not always interchangeable and we needed officers with special skills to deal with teenagers. We sought out volunteers for the school condition assignments. Officers who were actively engaged with neighborhood sports such as team coaches or were otherwise involved with scouts or other youth activities. We reasoned that these officers would be accustomed to dealing with rambunctious youths without over-reacting to them. We also required these officers to address students in classrooms or other assemblies so that the students would recognize the officers on the stations and the officers would get to know the students as well.

- As the result of an agreement with the P.B.A., field commanders were not made aware of civilian complaints unless the cases were substantiated. One of the difficulties in investigating citizen complaints, however, is that the cases usually involve only the complainant and the police officer. One person's word against the other. Without corroborating evidence those cases are routinely classified as unsubstantiated. Even when cases are substantiated, there is a substantial delay in notifying the field commanders due to the length of time needed for the investigation. We found that in some cases officers received several complaints while involved in a particular type of assignment such as plain clothes or school conditions. Because the field commanders were not aware of the existence of those complaints, the officers simply continued in those assignments.

  I met with the P.B.A. executive officers and discussed our plans to deal with the civilian complaint problem. We agreed that keeping field commanders abreast of those complaints as they were received would be in the best interest of the officers by avoiding circumstances that might contribute to those complaints. A reduction of complaints would avoid having to bring officers up on disciplinary charges.

Our overall objective of this program was the improvement of police services with the belief that a reduction of civilian complaints would result. After about 15 months, the rate of civilian complaints was reduced from over 100 per month to about 30 per month.

Walter Signorelli
NYPD Inspector (Ret.)

BIG CITY LEADERSHIP

For big-city police departments to effectively fulfill their mandates, leadership must come not only from the top, but also from every level within those departments. Individual leaders, including mid-management executives, first-line supervisors, and line officers, should have the capability to propose new initiatives and better solutions to the ever-evolving problems that police departments face. Unfortunately, in some cases, initiatives from the lower ranks are stymied by bureaucratic headwinds or obstructionists in the chain of command. During most of the thirty years, I served in the New York City Police Department, in ranks from police officer to Inspector; suggestions and proposals from the lower ranks were regularly ignored or shunted aside. It seemed that only initiatives emanating from headquarters were acceptable, and the workers who were required to implement the initiatives had little to say about the value of the initiatives or how they should be carried out. But there's always hope. In 1995 when I was the executive officer of the City-Wide Narcotics Division, I had the responsibility to supervise and monitor the effectiveness of our enforcement efforts. We had increased the number of personnel in the Narcotics Division from about 550 officers to almost 2000 officers, and commensurate with those additional resources, we increased arrests for narcotics violations more than 200%. However, the drug gangs that were plaguing the streets of many neighborhoods did not appear to be diminishing. For every arrest for narcotics we made and for every dealer taken off the street, as many new dealers seemed to appear. Nonetheless, the top echelon of the department continually demanded more and more arrests. But to many of us within the department who had battled against narcotics our whole careers, it was clear that simply arresting and re-arresting lower-level dealers would not solve the problem.

One morning, I suggested to my boss, the Chief of the Narcotics Division, that we should make fewer arrests, but more targeted arrests. We needed to arrest not just the street dealers but also the higher-ups, the suppliers, and the money men; but our practices and procedures were not conducive to making cases against people who did not "touch" the product. We needed to enlist the aid of all the pertinent law enforcement agencies to go after our targets through the use of tax laws, forfeiture proceedings, deportations, and license revocations. The idea was to get representatives from the DEA, FBI, IRS, Immigration, Customs, Secret Service, U.S. Marshals, State Police, and district attorneys to work under one roof, share information, and work on joint projects. This would foster personal relationships and the kind of trust needed for maximum effectiveness.

This was not an entirely new idea. It had come up before, but obstructionists had killed it, saying that it would never work because the "feds" would never share information or participate on equal terms. To my surprise, the Chief said let's do it; put a proposal together. We sent the proposal up the chain of command to Police Commissioner Howard Safir, and he enthusiastically approved it. Within six months, the "Police Commissioner's Manhattan North Narcotics Enforcement Initiative" was up and running. We got all the federal agencies on board, leased and renovated a four-story warehouse in Upper Manhattan, and moved in several hundred law enforcement personnel from the City, State, and federal government agencies. The results were truly impressive. Our joint investigations led to major arrests not only in New York but also across the country and in foreign countries. More importantly, we made a significant contribution to the stupendous crime reductions that occurred after 1995 and that have continued. For my part, it will always be gratifying that I was able to play a role in a successful initiative.

INDEX

A

Accomplishing stage, in leader management model, 25, 26
Accountability leadership, 185–187
Achievement-oriented style, 20, 112, 114, 120
Adaptation, 17
Adler, A., 131
Affect, teams and, 25
Ahear, L., 76
Almeida, C., 93
Alpert, B., 120
Ancona, D.G., 23
Andeweg, R.B., 131
Anticharismatic leadership, 164–165
Asimov, N., 74
Atwater, L.E., 13
Avolio, B.J., 50, 51, 53

B

Babwin, D., 62
Baker, T.E., 1
Barnett, J., 88, 89
Baruch, Y., 152
Bass, B.M., 8, 14, 16, 23, 24, 25, 50, 51, 53, 69, 70, 71, 81, 82, 84, 130, 132, 141, 151, 155
Bauer, T., 56, 57, 58
Beliefs, self-exploration of, 15
Belkin, D., 145, 147
Belkin, L., 38
Bellin, G., 132
Bennis, W.G., 8, 51, 51n, 152, 153
Berbeo, D., 91, 92
Bergner, L.L., 1
Berne, E., 10, 131
Bernstein, M., 87–89
Bias, leader-member exchange theory and, 36–37
Big five model of personality, 132
Big hairy audacious goals (BHAGs), 168–171
Birth order, 131
Black, C., 103
Blackmon, D.A., 56
Blake, R.R., 8, 68, 69, 71, 74
Blanchard, K.H., 9, 81–82, 84–86, 90, 119–122, 128
Bono, J.E., 52
Bostic, H., Jr., 76
Bragg, R., 103, 104
Branch, K., 103
Bratton, William, 112, 114–119, 122, 123
Brennan, C., 107
Brewer, N., 26
Bridges, T., 103
Broder, J.M., 118

Brooke, J., 106–108
Brown, Lee P., 37, 38–42
Brown, Willie, 73
Brubaker, P., 28
Bryant, A., 163, 164, 165
Bryman, A., 9, 52, 82, 100, 101, 113, 151, 161, 211
Bullitt, 132
Burns, J.M., 8, 13, 15, 16
Burritt, C., 109
Bush, George H. W., 85
Bush, George W., 126

C

Calder, B.J., 153, 154
Caldero, M., 16, 17
Caldwell, C., 15
Caldwell, D.F., 15, 16
Campbell, C., 121
Canellos, P.S., 146, 148
Cannon, L., 60
Cannon-Bowers, J.A., 24
Cartwright, D., 67
Casey, J., 75
Cashman, J.F., 8, 33, 34, 36
Casimir, L., 103
CEASE (Combined Enforcement and Supervision Effort) program, 76
Chacon, R., 146
Chadbourne, J., 82
Change intelligence (CQ), 142, 144
Charisma (charismatic leaders), 51–53
  need for, 52–53
  psychodynamic approach and, 133–138
Charlotte-Mecklenburg Police Department, 75–78
Charness, W., 143
Cherfils, M., 103
Christner, 69
Christopher Commission, 60, 61, 92, 94
Chronis, P., 106
Cincinnati, team policing in, 161
Ciulla, J.B., 12
Clawson, J., 137–138, 142, 144, 170, 171
Clines, F.X., 89
Coacting groups, 98
Cognitive processes, team theory and, 24
Cognitive resources theory, 100
Cohesion, team performance and, 25
Cohesiveness, 25
Colangelo, L., 136
Collins, Jim, 163–169
Combined Enforcement and Supervision Effort (CEASE) program, 76
Common goals, commitment to, transformational leadership and, 16

Community Mobilization Project (CMP), 59
Community Policing Consortium, 86
COMPSTAT (Command Accountability Strategies), 91, 117, 118, 119, 120, 164
Conceiving phase, in leader management model, 25
Conceptual skills, 1, 2, 10, 141
Conger, J., 51, 52
Connelly, M.S., 141, 143
Contingency contracting, 82
Contingency theory, 9, 97, 99–109.
  See also Vroom-Yetton contingency model
  case studies, 102–109
  criticisms of, 101
  dyadic approach to, 101
  military leaders and, 99–100
  overview of, 97–99
Contingent reward, 50
Coordination, team leaders and, 25
Corruption, 16
  definition of, 16
  economic, 17
Counteracting groups, 98
Craft ethics, 13
Crank, J.P., 16, 17
Critical Issues in Police Training (Haberfeld), 3, 170
Crothers, L., 2
Csikszentmihalyi, M., 170
Cullen, K., 145, 146
Curry, T., 61

D

Dainty, P., 69, 82
DARE (Drug Abuse Resistance Education), 59
Davis, P., 103
Dearborn, K., 144
Defining Issues Test (DIT), 13
De la Cruz, D., 136
DeMarco, P., 145, 147
Dessler, G., 112, 114
DeVanna, M.A., 52
De Vries, R.E., 51–53
Dinesh, R.M., 36
Directive style, 90, 112, 114, 119
Dobbins, G.H., 25
Doherty, M.L., 34
Dole, Elizabeth, 58
Domanick, J., 60
Donze, F., 121
Dotson, David, 61
Driscoll, A., 102
Drug Abuse Resistance Education (DARE), 59

Duchon, D., 36, 37
Duin, S., 88, 89
Dumaine, B., 23
Dunegan, K.J., 34, 35
Dussault, R., 115, 117
Dvir, T., 54
Dyads (dyadic relationships)
  antecedents to high-quality dyadic
    relationships, 36
  effectiveness and, 34–36
  in leader-member exchange (LMX)
    theory, 36–42

E

Early career leadership training, 1–2
Eckhart, D.E., 13, 16
Economic corruption, 17
Edwards, George, 151, 156–161
Effective leadership, six steps to, 137
Effectiveness, dyadic relationships
  and, 34–35
Emergent leadership, 26
Emotional affect, teams and, 25
Emotional intelligence (EQ),
  142–144
Emotional Quotient Inventory
  (EQ-i), 143
Epstein, G., 102
Erdogan, B., 34
Erez, A., 8, 26
Ericsson, K.A., 143
Estes, A., 146, 147
Ethics (ethical leaders)
  characteristics of ethical leaders,
    15–16
  craft, 13
  transactional leadership and, 53
  transformational leadership and,
    13–14
  under emphasis on, 12
  use of the word, 15
Evans, Paul F., 141, 144–148
Expectancy theory of motivation, 112
Extraversion, 132
Extroversion, 132

F

Fairness, of transformational
  leaders, 13
Feldman, D.A., 143
Fennell, E.C., 57
Fermino, J., 87
Fernandez, C.F., 81, 82
Fiedler, F.E., 98, 99, 100
Filosa, G., 121
First-line supervisors, 1, 12. *See also*
  Line officers
Fleischman, 70
Fleishman, E.A., 142
Flynn, K., 135
Follower motivation, 50
Foote, D., 60, 61

Ford, B., 147
Formal goals, 169
Freud, S., 10, 129, 130, 131, 132
Friedman, D., 61
Fuller, J.B., 53
Functional team leadership, 23–25

G

Garcia, J.E., 40, 97, 100
Gates, Daryl, 50, 55, 59, 60, 61, 62
George, A., 132
George, J., 132
George, T., 117
Georgopoulos, B.S., 112
Giampetro-Meyer, A., 13, 15, 51
Glionna, J.M., 72, 73, 74
Gobdel, B., 34
Goldstein, A., 59, 60
Goleman, D., 144
Gonzales, Elian, 5, 97, 101, 102
Grace, S., 121
Graeff, C.L., 82
Graen, G., 8, 33, 34, 36, 37
Grant, J.D., 26
"Great Man" theory, 10, 151
Greenberg, Reuben, 50, 55–58
Greene, J.A., 116, 117
Greenleaf, R., 15
Greimel, 87
*Group Psychology and the Analysis of
  the Ego* (Freud), 129
Guy, Robert, 76

H

Haberfeld, M.R., 170
Hahn, James K., 93, 118
Hall, B., 169
Hallinan, Terence, 73
Halpin, 69, 70
Hardin, J., 57
Harman, David, 27, 28, 29, 30, 31
Harrell, J., 28, 29
Harvard University, leadership-style
  studies, 68
Hays, T., 134, 136
Heaney, J., 147
Hedgehog rule of management,
  168
Heier, W.D., 97, 99, 100
Heifetz, R.A., 14
Hemphill, 69
Hersey, P., 9, 81, 82, 84, 85, 86, 90
Hicks, B., 56
Hill, W., 97, 98
Hoegel, M., 26
Hogan, R., 132
Hollander, E.P., 151
Holly, D., 89
Honesty, 16
House, R.J., 8, 9, 51, 52, 54,
  113, 114
Hutchinson, E.O., 92

I

Idealized influence, 51, 53
Individualized consideration, 51, 53
Indvik, J., 112
Informal goals, 168, 169
In-groups, leader-member exchange
  (LMX) theory and, 33, 34, 35, 36
Inspirational leadership, 51
Integration of behavior and values, 14
Integrity, 12–31
  definition of, 12
  five-step approach to
    management of, 21
  recruitment and, 16–17
  selection and, 17–18
  training and, 18–19
  triangle of, 16–19
Intellectual intelligence (IQ), 142, 144
Intellectual stimulation, 51, 53
Intelligence
  change (CQ), 142
  emotional, 142
  emotional (EQ), 144
  intellectual (IQ), 142
  social, 143
  social (SQ), 142
Interacting groups, 98
Introjection, 130
Ippolito, M., 122
Isenberg, D.J., 25
Israeli Defense Forces, 54

J

Jacobs, D., 57
Jacobson, M., 116, 117, 118
Jago, A.G., 151, 152, 153
Janis, I.L., 24
Jenkins, 154
Jermier, J., 9, 100, 112, 113
Jessup, H.R., 26
Jimenez, J.L., 103, 104
Judge, T.A., 52
Jung, C., 10

K

Kandel, J., 118
Kanungo, R.N., 14, 16, 52, 55
Kastenbaum, D.R., 98
Katz, C.M., 59, 60, 116, 117
Kennedy, H., 89
Kerik, Bernard, 129, 133–138
Kerr, N.L., 26
Kerr, S., 9, 100
Kets de Vries, M.F.R., 130
Kidwell, D., 103
Kiker, S., 9, 100
King, Rodney, 39, 60–62
Kirkpatrick, S.A., 10, 153
Kirmeyer, 70
Klockars, C.B., 12, 16, 21
Kobe, L.M., 10, 143

Koby, Tom, 105–109
Kohut, H., 130
Korten, D.C., 83
Kozlowski, S.W.J., 24, 25, 34
Kramer, M., 60
Kranes, M., 87
Krimmel, J.T., 155, 156
Kroeger, O., 10, 132
Kropf, S., 58
Kuehl, C.R., 97
Kuhnert, K.W., 8, 13, 16, 53, 55
Kuykendall, J., 84, 85

L

LaBarre, W., 1300
Laird, L., 118
Laissez-faire management, 50
Lakshmanan, A.R., 145
Lakshmanan, I., 116
Larson, J.R., Jr., 24
Leader Behavior Description
    Questionnaire (LBDQ),
    69, 70, 112n
Leader-management model,
    25–26
Leader Match program, 98–99
Leader Member Exchange (LMX)
    instrument, 8, 34
Leader-Member Exchange (LMX)
    theory, 8, 33, 34, 35, 36
    antecedents to high-quality dyadic
        relationships, 36
    bias and, 36–37
    effectiveness and the dyadic
        relationship, 34–35
Leadership (leadership skills).
    anticharismatic, 164–165
    choice of the right people and,
        167–168
    definition of, 3
    emergent, 26
    hedgehog rule of management and,
        167–168
    Level 5, 165–167
    pentagon of, 5–10, 19–21, 26–29
    of public service workers, 2
    team. See Teams
    theories of, 3
Leadership Effectiveness and
    Adaptability Description (LEAD)
    instrument, 82, 85
Leadership theories, 8–10
Leadership training
    advanced, 7
    early, 1–2
    gap in, 3–5
Leader substitutes theory, 100
Learned leadership skills, 141–142
LeBlanc, P.M., 33, 34
Legitimate authority, bases of, 129
Lelchuk, I., 72
Lenzner, R., 163, 168

Leovy, J., 118
Lessem, R., 152
Level 5 leadership, 165–167
Lewin, K., 98
Lewis, D.L., 117, 118
Lewis, P., 13, 16, 53, 55
Liden, R.C., 8, 33, 34, 36
Lin, 70
Lindenmuth, P., 155, 156
Line officers, 1–2, 4, 6
Liska, L.Z., 112n, 113
LMX (leader-member exchange)
    theory, 33–46
    antecedents to high-quality dyadic
        relationships, 36
    bias and, 36–37
    effectiveness and the dyadic
        relationship, 34–35
Locke, E.A., 10, 152, 153
Lord, R.G., 10, 143, 154
Los Angeles Police Department
    (LAPD), 59–62, 91–94, 118
    Rampart scandal, 92–93, 118
    team policing in, 26–27
Los Angeles Police Protective League
    (LAPPL), 92
Lowe, 53
Lundy, S., 57
Lying, 16

M

McCanse, A.A., 68, 69
McCauley, L., 56
McCormack, F., 57
McCoy, M.C., 25
McCullen, K., 105–109
McGreevy, P., 91, 92, 119
McGrory, B., 116, 146
Macht, J., 165, 166
McPhee, M., 105–107
McQuillan, A., 117
Malkin, M., 89
Mallia, J., 145
Management-by-exception, 8, 50
Management Leadership
    Questionnaire (MLQ)
    instrument, 53
Managerial grid, 68–69
Mann, R.D., 10, 133, 154
Manning, 89, 90
Maple, Jack, 118
Marantz, S., 147
Martinez, J., 147
Martinez-Carbonell, K., 12
Marzulli, J., 116, 133–135
Masi, R.J., 54
Mathieu, J.E., 24, 114
Matier, P., 74
Maurer, T.J., 35
Menchaca, R., 57
Menendez, J., 146
Messing, P., 135, 136

Michigan University leadership
    studies, 66–68
Middle management, 1, 2
Military leaders
    contingency theory and, 99–100
    style theory and, 69–70
Misumi, J., 25, 66, 69
Montclair Police Department
    (New Jersey), 27–31
Moose, Charles A., 58, 81, 86,
    87–90
Moral altruism, 14
Moral development, 50
Moral reasoning, 13, 14
    transformational theory and, 53
Morgan, S., 91
Morse, R., 74
Moses and Monotheism (Freud), 130
Moulton, J.S., 8, 69, 71, 74
Mullener, E., 121
Mumford, M.D., 141, 142,
    143, 144
Munday, D., 56
Myers-Briggs Personality Profile, 132

N

Nahavandi, A., 26, 27
Nanus, B., 8, 51, 51n, 152, 153
Narcissism, 130
Nash, J., 118, 119
National Commission on Terrorist
    Attacks Upon the United States
    (9-11 Commission), 135
Negri, 145
Newfield, J., 116, 117, 118
New Orleans Police Department, 42,
    120–124
New York Police Department
    (NYPD), 114–119, 133
Niehouse, O.I., 83
Niesse, M., 121, 122
9-11 Commission, 135
Nkrumah, W., 87
Noble cause corruption, 16, 17, 19
Noonan, E., 146
Northouse, P.G., 3, 8, 9, 10, 34,
    66, 67, 68, 69, 81, 112, 137,
    151, 154
Nowicki, Dennis, 71, 75–77
Nurturing, ethical leaders as, 15

O

Objective needs for leadership, 53
O'Brien, William, 101–105
Offer, D., 130
Offermann, L.R., 151
Ohio State University, leadership
    studies at, 66
Open communication leadership,
    175–176
Operation Scrap Iron, 146

Organizing phase, in leader
  management model, 25
Orlov, R., 60, 61, 62, 93
Oshiro, G.R., 88
Owens, J., 152

P

Parker, William H., 59
Parks, Bernard, 58, 81, 86, 91–94
Participative leadership, 67–68
Participatory leadership, 24
Path-goal theory, 9, 20, 21
  case studies, 115–124
  military leadership and, 114
  overview of, 112–114
Pennington, Richard, 112, 114,
  120–124
Pentagon of police leadership, 5–10
Perlstein, M., 120, 121
Pernick, R., 141, 142
Peters, L.H., 97
Piacente, S., 57
Pickard, J., 164, 166
Plutarch, 151
Police integrity. *See* Integrity
Police leadership. *See* Leadership
Porras, J., 168, 169
Port, B., 136
Porter, A., 58
Price, T., 13, 14
Proactive training, 3–4
Psychodynamic approach, 129–138
  birth order and, 131
  case study, 133–138
    to Commissioner Bernard Kerik,
    133–138
  overview of, 129–131
  personality and, 132–133
  psychohistory and, 132
  transactional analysis and,
    131–132
Psychohistory, 132
Public service workers, leadership
  skills of, 2
Punch, M., 16
Pyle, R., 134, 136, 138

Q

Quaintance, M.K., 142
Quality of police leadership, 1

R

Rampart scandal, 92, 93, 118
Ramsey, Jonbenet, 105–109
Ranalli, R., 146
Rashbaum, W.K., 134
Reactive training, 3
Recruitment, 7
  integrity and, 16–18
Reddin, W.J., 68, 81

Reiterman, T., 72–74
Relationship-oriented leaders, 98
Rennie, M.W., 132
Rewards, situational leadership
  theory and, 83
Rice, R.W., 98
Richards, D., 114, 115
Robey, R., 105, 108
Rogers, J., 42, 44, 45, 92
Rosenfeld, S., 74
Rosenwald, M.S., 146–147
Rotated leadership, 26
Rowley, J., 152
Rucker, A., 57

S

Saha, S.K., 97
St. Bernard, S.C., 120
Salas, E., 23
Sanders, Earl, 66, 71–78
Sarachek, B., 151
Scandura, T., 36
Schodolski, V.J., 92
Schriesheim, C.A., 33, 97, 99, 112
Selection of personnel integrity
  and, 16
Self-exploration of beliefs, values,
  and assumptions, stewardship
  and, 15
Self-managed teams, 26
Self-management, 24
Sennott, C.M., 147
September 11, 2001, terrorist attacks,
  134–138
Servant leadership, 15
Service-oriented ethical leaders,
  15–16
Shah, D.K., 59
Shamir, B., 52, 54
Shapiro, J.S., 59, 61
Shenon, P., 135
Sheriff, D.R., 10, 141
Siemaszko, C., 117, 118
Sisk, R., 87, 89
Situational leadership theory, 81–94
  case studies, 87–94
  employee burnout and, 83
  overview of, 81–83
  policing and, 84–86
  politics and, 83–84
  simplified version of, 83
Skills approach, 3, 141–148
  case study, 145–148
  leadership in context and, 142–143
  learned leadership skills, 141–142
  overview of, 141
  U.S. Army studies, 144
Social intelligence (SQ), 142, 144
Sperry, P., 89
Standora, L., 135

Stech, E.L., 129, 131, 132
Steidlmeier, P., 14, 16
Steiner, D.D., 36, 37
Stephens, Darrel, 37, 42–46
Sterngold, J., 93
Stevens, K., 56
Stewardship model, 15
Stewart, J., 91
Stirgus, E., 120–121
Stogdill, R.M., 8, 10, 69, 70, 81, 151,
  152, 154, 155
Stolberg, M.M., 157–160
Stone, K., 59, 60
Stress, teams and, 25
Strozier, C., 130
Strube, M.J., 97
Style theory, 66–78
  case studies, 72–78
  Harvard University studies, 68
    managerial grid and, 68–69
  Michigan University leadership
    studies, 66–68
  military leadership and, 69–70
  Ohio State University leadership
    studies, 66
  overview of, 66
  police leadership and, 70–71
  as a universal theory, 69
Subjective needs for leadership, 53
Suggs, E., 120–122
Sugiman, T., 25
Supervisory planes, 1–2
Supportive leadership, 113
Swanson, C.R., 1, 2
Sward, S., 72–74
Szaniszlo, M., 147

T

Tannenbaum, S.I., 24
Task structure and leaders, 99
Task-oriented leaders, 98
Team policing, 26–31
Teams
  effective team leaders, 23–24
  functional team leadership, 23
Teams (team theory; team
  leadership), 23–31
  affect and, 25
  cohesion and, 25
  definition of, 23
  functional team leadership, 23–25
  leader-management model, 35
  self-managed teams, 36
Terzian, P., 89, 90
Tesluk, P.E., 24
Tharp, M., 61
Theories of leadership, 7, 67
Theusen, J.M., 10, 132
Thorbourne, K., 29
Thorndike, E.L., 143

Tichy, N.M., 52
Toche, H., 26
Top management, 2
Topousis, T., 135
Tosi, H.L., 9, 100
Training
  early, 1–2
  gap in, 3–5
  integrity and, 18–19
  proactive, 3–4
  reactive, 3
Traits, 10
Trait theory, 151–161
  case study, 156–161
  modern applications of, 152–154
  overview of, 151–152
  social scientific evaluation of
    leadership traits, 154–156
Transactional analysis, 131–132
Transactional leaders, 8
Transactional leadership,
  transformational leadership and,
    13–14, 55
Transformational theory
  (transformational leadership),
    13, 50–65
  charisma and, 51–53
  ethics and, 13–14
  factors in, 50
  meta-analytic studies on, 53
  in military contexts, 53–54
  moral reasoning and, 53
  transactional leadership and,
    13–14, 55

Tsao, E., 89
Tucker, C., 122
Turner, N., 13, 51, 62, 123
Turque, B., 60, 61
Typewatching, 132

U
Uhl-Bien, M., 8, 33, 37
Unsinger, P.C., 84, 85
Utecht, R.E., 97, 99, 100

V
Values
  integration of behaviour and, 14
  self-exploration of, 15
Van Auken, P., 83
Van Den Berg, S.B., 131
Van Derbeken, J., 72–74
Varney, J., 120
Vecchio, R.P., 34, 81, 82, 100, 101
Vertical linkages, in leadership, 33
Villegas, M., 93
Vinzant, J.C., 2
Viturello, Frank, 23, 27–31
Von Glinow, M.A., 112
Vroom, V.H., 9, 100–102, 112
Vroom–Yetton contingency model, 9,
  100–102

W
Walker, S., 58, 59, 60, 116, 117
Weber, J., 14
Weber, Max, 51, 129
Webster, William, 61

Weinkauf, K., 26
Weiss, M., 135, 136
Weller, R., 108
Werner, E., 88, 89
West Point Military Academy, 3
Whyte, David, 130, 131
Wilborn, P., 91, 92, 94, 118
Williams, Willie, 61
Wilson, Woodrow, 132
Winer, J.A., 130
Wisby, G., 75
Wofford, J.C., 112n, 113
Woodall, B., 76
Woodlief, W., 147
Woodrow Wilson's idealization, 132
Work groups, 98

Y
Yammarino, F.J., 34, 53, 143
Yang, J.E., 61
Yetton, P.W., 9, 100, 102
Young, T., 121
Yrle, A.C., 35, 37
Yukl, G., 9, 66, 67, 69, 81, 82, 100, 113

Z
Zaccaro, S.J., 8, 23–25, 141, 143
Zadrozny, A., 28
Zaleznik, A., 130, 132,
  151, 152, 153
Zamora, J.H., 72
Zander, A., 67
Zarella, J., 103
Zelditch, M., 68